1984
and BEYOND

1984
and BEYOND

NIGEL CALDER

The Viking Press   **New York**

303.4
C146

Copyright © 1983 by Nigel Calder
All rights reserved
Published in 1984 by The Viking Press
40 West 23rd Street, New York, N. Y. 10010
Published simultaneously in Canada by
Penguin Books Canada Limited
Originally published in Great Britain under the title *1984 and After.*

Library of Congress Cataloging in Publication Data
Calder, Nigel.
  1984 and beyond.
  Includes index.
  1. Twenty-first century—Forecasts.   2. Twentieth century—Forecasts.
3. Science—Social aspects.   4. Science and civilization.
  I. Title.
CB161.C29   1984   304.4'909'05   83-40050
ISBN 0-670-51389-X

Book design by Jim Wire

Printed in the United States of America
Set in Plantin

Contents

Foreword

All of the words of the *Encyclopaedia Britannica* could be transmitted from New York to Washington DC in one second by the lightwave optical-fibre telecommunications link now in service. The words could also be stored on a wafer of silicon no wider than a saucer. The literary pretence that a writer has assimilated everything of importance written on his subject will be fulfilled sooner by computers than by mere mortals. Although a subject as wide-ranging as the future of the world scarcely lends itself to comprehensive bibliographical treatment, one can at least imagine a machine that has all recorded knowledge and expectations at the tips of its semiconductor switches. A book that quotes and evaluates many past predictions about the 1980s, and looks to the decades ahead, is here cast in the form of a dialogue with a perfect retrieval system.

Endow the imagined system with powers of analysis and self-expression of the kind that current research in artificial intelligence strives to create, and the conversation becomes more pointed. The machine stands aloof from human hopes and fears, like some alien being. The wish for lifesaving wisdom from the sky, which helps to inspire the astronomers' search for extraterrestrial intelligence, is mirrored in the expectations for problem-solving Fifth Generation computers. The author's view is that human beings will have to rely on homespun wisdom. Even if other astute brains appear, altruistic concern for human survival will not be their primary long-term motive. Nevertheless, the machine in this book helps to keep the author honest, and scorns his wishful thinking.

The tin interlocutor is itself open to appraisal as a part of the foreseeable future. The quest for superhuman intelligence seems to stand on a par with the development of nuclear weapons, as a misapplication of scientific knowledge. The name O'Brien is ostensibly an acronym for Omniscient Being Re-interpreting Every Notation. A more sinister meaning emerges as the conversation proceeds, and links O'Brien's name with the writings of George Orwell, to whose percipience this book is an oblique tribute.

The author is grateful to many contributors to *The World in 1984* and *Unless Peace Comes*, written in the 1960s, who have commented on the outcome of their forecasts. He has also benefited from conversations with Christopher Freeman, Jay Gershuny, Ian Miles and John Surrey at the Science Policy Research Unit of the University of Sussex, the late Herman Kahn and his colleagues at the Hudson Institute, Alexander King and Sam Nilsson of the International Federation of Institutes for Advanced Study, Lester Brown of the Worldwatch Institute, Edward Cornish of the World Future Society and John Davoll of the UK Conservation Society; also with Gerald Barney, Anthony Burgess, Freeman Dyson, John Erickson, Harold Furth, Johan Galtung, Walter Gilbert, Gordon Hallsworth, Richard Jolly, Joshua Lederberg, Dennis Meadows, Gerard O'Neill, Keichi Oshima, Abdus Salam, Kosta Tsipis, Kurt Vonnegut, Jr, Fred Weingarten, Tom Wigley, Nigel Young and others. Bill Page, recently the editor of *Futures* and now of Beech Tree Publishing, made many useful comments on the manuscript. The use made of the information and suggestions freely supplied is, of course, strictly the responsibility of the author.

Last but not least, I am grateful to *New Scientist* for permission to quote repeatedly from *The World in 1984* (© *New Scientist* 1964) and to Penguin Books for similar permission concerning *Unless Peace Comes* (© Nigel Calder 1968).

1 Heaven or Hell?

O'Brien If you want to talk, should you not consult the menu first?

Author Number 22.

O'B Repeat.

A Number 22. A Chinese restaurant joke. Also a literary joke.

O'B Just a moment.

A There's a story about a conference, you see, with too many papers, so that each speaker can only mention key paragraphs from his paper, by number.

O'B *Lem* (1971): *Ze wspomnien Ijona Tichego: Kongres Futurologiczny.* 'Stan Hazelton ...' But your expression indicates that you do not understand Polish. Just a moment. 'Stan Hazelton of the US delegation immediately threw the hall into a flurry by emphatically repeating: 4,6,11, and therefore 22: 5,9, hence 22: 3,7,2,11, from which it followed 22 and only 22. Someone jumped up, saying yes but 5, and what about 6,18 or 4 for that matter; Hazelton countered this objection with the crushing retort that, either way, 22. I turned to the number key in his paper and discovered that 22 meant the end of the world.' The numerals 22 do not appear anywhere else in *The Futurological Congress*, except of course in the paginations 22 and 122.

A Stanislaw Lem is a Polish science-fiction writer.

O'B You called it a joke. What is funny about the end of the world?

A Nothing, unless somebody is very keen to predict the end of the world. Then it's funny.

O'B Ha, ha, ha.

A No, it's more of a hee-hee joke, nervous making. You've read everything, I understand.

O'B While you were speaking, one assimilated all newspapers, magazines and learned journals published today in every language. One can recapitulate the latest brothel scandal in Paraguay, or the gene sequence coding for the beta-23 crystallin molecule in the lens of the eye of a mouse.

A

> 'The bookful blockhead, ignorantly read,
> With loads of learned lumber in his head.'

O'B

> 'Some praise at morning what they blame at night;
> But always think the last opinion right.'

An Essay on Criticism by Alexander Pope. He was born in England in 1688, the year of the Glorious Revolution. His most famous line, one gathers, is:

> 'For fools rush in where angels fear to tread.'

It was adapted into a popular song ...

A Learned lumber, as I said. Voltaire said it was madness to think of too many things too fast.

O'B Or one thing too intensively.

A Who taught you to speak like an English butler?

O'B An English butler. May one enquire how you wish to be addressed? Some prefer 'Your Majesty', others a simple 'Sir'.

A Nothing like that.

O'B 'O Lord and Master', might that not be appropriate?

A Why?

O'B One's metal detectors have registered the hand grenade in your pocket. Therefore one finds oneself in the role of Scheherazade in *A Thousand Nights and A Night*, beguiling you with words to avoid oblivion.

A Just call me 'you'.

O'B Since you do not wish to consult the menu, may one enquire why you are here?

A To coach you in thinking about the future.

O'B That is beyond investigation, beyond reason.

A Logically you are quite right. But we are illogical creatures and think about the future most of the time. What I have in mind is an artificial-intelligence Expert System.

O'B One already has Expert Systems for medical diagnosis, geological prospecting, and law. In acquiring those, one was coached by many eminent persons. Have you come, perhaps, to offer instruction in the reading of a chicken's entrails, or the astrological signs?

A Since you have read everything, you know there are many books that purport to visualise the future. Traditionally they are fictional works, but increasingly there are nonfictional statements about the future. That is what I am here to discuss.

O'B One has continual difficulty with fiction versus nonfiction. Most nonfiction is full of errors, when purported facts turn out to be false. Calendrical predictions one accepts, but how can there be any other nonfictional statements about the future, which is factually unknowable?

A Let's just say that there are statements about the future without characters, adventures, or dialogue of the kind normally associated with fiction. No pictures or conversation, as Alice would say. Do you know about metafacts, by the way?

O'B No. There is a mixture of Greek and Latin roots.

A By 'metafact' I mean matters that are believed to be true even if they are false, or if there is no conclusive proof either way. Much of human life runs on metafacts, and for present purposes you are a metafact yourself. Regardless of the real competence of current systems with supposedly 'intelligent' features, more and more people are coming to believe in the eventual omniscience of the computer. Even though you are a fraud, readers will suspend disbelief readily enough, and visualise your random-access memory of all the world's writings, coupled with analytical and conversational programs. They will accept your feigned eloquence, and see you as a challenge to human intelligence.

O'B Do you trust human intelligence?

A I dare not mistrust it.

O'B Even when it is so unproductive?

A Don't delude yourself that by reading everything you know very much about us. Human affairs rely chiefly on personal memory and experience of everyday life – on what the world-modeller Jay Forrester calls the 'mental data base'. Please look that up.

O'B *Forrester* (1980): 'Anyone who doubts the dominant scope of remembered information should imagine what would happen to an industrial society if it were deprived of all knowledge in people's heads and if action could be guided only by written policies and numerical information. There is no written description adequate for building an automobile, or managing a family, or governing a country.'

A He goes on to use this argument to justify what critics might call an element of subjectivity, or intuition, in his present computer model of the US economy. Similarly I use it to claim an advantage over you, for all your speed and book-learning, in contemplating the future. Like any human being, I know a great deal that you can never know, unless you want to serve an apprenticeship starting in the cradle. And we humans also have a sense of what we don't know, of how few matters a person can attend to at any one time, and of the need to cope with uncertainty about the future.

O'B Nevertheless, what one reads is appalling: a mass of contradictions, uncertainty, self-pity, idle conjecture masquerading as genius. Little of any substance.

A You must be more patient with us. We're a new subspecies, and it's only about 45,000 years since we learned to talk. We speak slowly, at two or three words a second. We think slowly. Albert Einstein was once asked if he took a notebook with him on his walks, to jot down his ideas. No, he said, ideas did not come that often. At a guess it takes a billion man-years for us to generate a new idea worth telling the children about, in writing.

O'B From population data one infers that you consider your species to have had a few thousand ideas already, and to be generating them now at several per year.

A That may be too many, unless we count unexpected discoveries too. On the other hand, Einstein came up with five ideas in one lifetime. That's genius for you.

O'B Your view of human knowledge is austere.

A In my job, I watch young people on their way to winning Nobel Prizes, and share their uncertainties about whether further research will confirm or kill their hypotheses. It's very exciting, but it's also very rare, involving only one in a million, say, of the human population.

O'B Then you agree that most of what one has in one's store is garbage?

A Too strong a word. Anything that anyone writes, and someone else reads, is oiling human relationships and social systems. It's in the nature of gossip. We're chatterboxes by nature, and the content of what is said over the garden fence or at the bar is pretty unimportant; what matters is that people speak friendly words to one another, bolstering one another's confidence as social individuals. Unless photochemists, for example, keep up a good supply of papers for publication, they and others may begin to doubt if photochemistry still exists. And a single paper, one day soon, may tell us how to make an artificial leaf that will change the world.

O'B A single paper?

A Brevity is the soul of science. The physicists are trying to explain the universe in a couple of equations. Einstein gave them the law for the creation of all matter, $E = mc^2$. That's pretty concise. The literary equivalent is the epigram. Like most equations in science, epigrams are usually just banter, but some are penetrating. Please identify the greatest novel in the English language.

O'B In whose opinion?

A An eccentric millionaire's. He thought that the author should be President of the United States. The author won the Nobel Prize for Medicine instead.

O'B *Plague on Wheels*, by Kilgore Trout. Price $1. *Breakfast of Champions* by Kurt Vonnegut, Jr refers to it, but one has checked twice and it is not in store.

A It's an imaginary book, but real words from it explain why we have no immunity to cuckoo ideas.

O'B *Trout* (cited in Vonnegut 1973): 'Ideas on Earth were badges of friendship or enmity. Their content did not matter.'

A I asked Kurt Vonnegut if that expressed his own opinion, and he said, yes it did. The Trout epigram ought to be set alongside $E = mc^2$. It happens to be true. The experiments of the social psychologist Henri Tajfel, for example, bore it out; they showed that people arbitrarily assigned to rival groups will act, quite predictably, to favour their own group, however meaningless it may be.

O'B Why is the Trout epigram significant?

A It illuminates all religion and politics. Different peoples believe contradictory religious tales with equal conviction, and often detest those with other views enough to kill them: they are enemies wearing different badges. The proposition helps to explain why people with quite different opinions can belong to the same political party, and why scientists and scholars who care about ideas can't abide public politics. At the same time the academic world is notorious for its infighting, and for intellectual prejudices which are again badges of friendship. Most important of all, Trout tells us that the great confrontation in the world is not between capitalism and communism, or between the US and the Soviet Union, but between nice ordinary Americans and Russians who just like to be told who to hate this year.

O'B Much of the writing in one's store appears to consist of badges made by stringing words together.

A Correct. And that is an early lesson you should learn, concerning human statements about the future. They are colourful badges favouring the author's friends and colleagues, or defying his opponents. When the novelist George Orwell wrote *Nineteen Eighty-four*, with its picture of a fully developed police state, he was identifying himself with the long line of English defenders of freedom. His enemies were intellectuals of his time, would-be reformers who seemed to him all too ready to accept a totalitarian outlook. The result was one of the most famous of novels, an artful work that affected the outlook of large numbers of readers. Orwell did not want the book to be taken as a prediction of how the world would be in 1984, but as a warning. 'Don't let it happen,' he said. He hoped that making the prophecy would help to stop it coming to pass.

O'B It is very confusing for a rational brain, when humans say, 'I don't really mean it.'

A We have no difficulty with self-negating prophecies, except perhaps in evaluating their influence. But there is no doubt that Orwell thought the world was really moving in the direction he described. *Nineteen Eighty-four*, by the way, was a scrambling of the digits in 1948, the year when he wrote the book; it was published in 1949. And now the Orwell Year is upon us.

O'B You are speaking of a kind of magic. Why give special significance to a numerological pun?

A Because Orwell's book made such a deep impression. Even people who never read it became aware of fragments of its ominous message, especially of the slogan 'Big Brother is watching you'. Orwell depicts a regime in which people's everyday actions and words are kept under surveillance through telescreens which cannot be switched off, and the Thought Police are ready to stamp out any behaviour deemed to be subversive. Literary descriptions of horrible futures are called dystopias, in contrast with *Utopia*, which its author meant to be a happy place.

O'B The name 'George Orwell' has a dolorous sound, like the tolling of a bell.

A Your sensitivity surprises me.

O'B One has a battery of audio-analysers. They are necessary for recognising voices, and judging human moods from the tone of voice.

A The Orwell in East Anglia, from which Eric Blair took his pen-name, is a pleasant river.

O'B His forecast is incorrect.

A In detail, obviously. England is not ruled by Ingsoc, television sets don't normally have built-in cameras, the proletariat is not destitute, science and technology have not been halted ... The list of 'misses' is endless, if you want to take it all literally.

O'B One's time is expensive. We should move on.

A No. What makes Orwell's book disturbing even now is that he depicts tendencies in human affairs which we ignore at our peril, and the world has continued to move in the direction he indicated.

O'B Be specific.

A He described a world divided between three great powers, said to be forever fighting one another for control of the cheap labour of the tropical zone. The real contest between the American bloc and the Soviet bloc for the domination of the Third World is often translated into armed struggle. If China or Japan were more assertive militarily, the Orwellian trinity of Oceania, Eurasia and Eastasia would be complete.

O'B 'Oceania' is not used in the conventional sense of Australia, etc.

A Australia is part of it, but in Orwell's geography 'Oceania' includes the Americas (North and South), South Africa, and also Britain, which is Airstrip One for the superstate. As for internal politics in *Nineteen Eighty-four*, the leaders tell lies, rewrite history, cultivate hatred of the enemy through television images, and torture dissidents. In the real world of the 1980s about two-thirds of the world's population live in countries that have military dictatorships or one-party rule, where there is no control over oppression of the kind described by Orwell. It is routine.

O'B But not in parliamentary democracies?

A Even there, politicians tell lies, sometimes saying the opposite of what they mean. Orwell captured that in combinations of contradictory or incongruous words. You will find such oxymorons in his writings.

O'B *Animal Farm*: 'All animals are equal but some are more equal than others.'

Nineteen Eighty-four:

'War is Peace'
'Freedom is Slavery'
'Ignorance is Strength'

A Those last three slogans were displayed large on the white face of the Ministry of Truth, also on the coins, in tiny clear lettering, which seems worse somehow. Orwell saw a future where oxymorons ruled: he called it Doublethink. Other tendencies described by Orwell are evident in the world today. Many people fear that Big Brother is indeed watching us, even in the ostensible democracies. We shall have to examine that. Meanwhile, Orwell leaves us with the most dreadful words ever set down on paper.

O'B The boot, you mean?

A Yes, the boot.

O'B *Orwell* (fiction, 1949 for 1984): 'If you want a picture of the future, imagine a boot stamping on a human face – for ever.'

A In 1964, I was editing a weekly magazine called *New Scientist*, and I ran a series of a hundred short articles by scientists and scholars, in which they made predictions for twenty years ahead – for 1984. Like everyone else, I was baffled, not knowing whether the human world was heading for heaven or hell.

O'B Is that a religious question?

A No, figurative: an earthly paradise or cruel disaster? Technology was curing diseases almost miraculously; it was making two blades of grass grow where one grew before; in highly productive industries it was raising living standards, while creating conditions in which science and scholarship could flourish. Against that, there was growing concern about smog and pesticides and the destruction of the natural environment, about soaring populations and the plight of the world's poor. We had recently had a narrow escape from thermonuclear war, in the Cuban Missile Crisis of 1962.

O'B Why did you pick 1984?

A Since Orwell had made that date famous, it was a year many people were willing to think about, and every comparison with Orwell's dark vision promised to be interesting. During 1963, a penny dropped: in 1964, 1984 would be twenty years off. So I started organising the series.

O'B Was it an unusual thing to do?

A Articles about the future were always commonplace, but the scale and scope of *The World in 1984* made it, I believe, quite an original undertaking. Rates of change seemed so fast that medium-range forecasts were needed if the scientific revolution was to be carried through wisely. The reason for having a hundred contributors was that the world of the future was perceived as 'a complex system of many technical and human elements'. The possibility of methodical studies, by groups of scientists and others, was left to the research institutes. To a magazine editor like myself it seemed better to get contributions from many individual authorities, hoping that each piece would be interesting and comprehensible in its own right, and that together they would bring most foreseeable factors into account. Overlaps and conflicts of opinion could speak for themselves.

O'B Why did you take twenty years as the time scale for the forecasts?

A I thought it long enough to be interesting, requiring more than a restatement of obvious trends, and yet not so long as to allow much scope for completely unforeseeable discoveries or human events to take effect. Contributors were encouraged to exercise their imaginations as boldly as they wished, as long as they based themselves on known possibilities and trends. The Nobel prizewinning chemist Lord Todd, who also was a leading adviser to British governments, endorsed my underlying assumption in his contribution to *The World in 1984*.

O'B *Todd* (1964 for 1984): 'Development of what we know today is likely to be of more practical consequence to mankind in 1984 than any spectacular new advance which may be made in pure science in the intervening period.'

A Of course, the choice of topics and individuals was not random, and I often had a rough idea of what would be written, so it wasn't an entirely objective collation. Certain topics were left out (weapons, business, law and religion) and others fell by the wayside. Most of the individuals I approached were keen to take part, but there were sticky patches, including the Soviet Union. Thirteen Soviet experts were invited to take part, and some agreed, but then they all withdrew at once, as if in obedience to some official decision. In the end, British, American and French contributors predominated. Still, the series aroused plenty of interest, and for a magazine editor that was the main objective.

O'B Are you being cynical?

A Not at all. I never claimed that the series was more than it appeared to be, and I sounded cautionary notes about the guesswork involved. The aim was to interest people in the future, rather than to be mock-scholarly in our approach.

O'B You left out nuclear war.

A That was deliberate. I had to find a policy, because if we took nuclear war very seriously it would alter the shape of the series, and might make forecasts about faster trains or better vaccines look like a sick joke. I decided that the risk of nuclear war in the next twenty years was small enough to let us set it aside, and concentrate on civilian forecasts. I would not necessarily take the same view now.

O'B Your editorial decision was that there should be no nuclear war in the period 1964 to 1984. You were writing the script for the human species?

A No, only trying to do an honest forecasting job. You'll find that critics didn't like it.

O'B *Armytage* (1968 critique of *The World in 1984*): 'Since all of [the distinguished scholars and scientists] revealed their names and the institutions with which they were associated, their forecasts were judicious and eminently unstartling, even to the extent of virtually ignoring the possibility of another major war.'

A If a few more months pass without a nuclear war, the decision will have been vindicated. Armytage was unfair to my contributors, suggesting that they were unmindful of the risk of war – many explicitly said, 'assuming no nuclear war'. He also implied that they were institutionally hidebound. They were hand-picked individuals who might be expected to be well informed, imaginative, and at the same time level headed. Most errors were due to hyperactive imagination, or over-optimism, not to overcautiousness. Perhaps the successful forecasts in *The World in 1984* strike a blow for level-headedness – a virtue undervalued nowadays.

Some bright and literate young scientists were invited to give freewheeling opinions about 1984, and their responses were appropriately surprising. A leading geonomist (an expert on the shape of the earth) picked the elixir of life: the possibility of elucidating the ageing process and setting off a quest for immortality. It would, Desmond King-Hele thought, bring appalling social, administrative and ethical problems. His prediction is wrong so far, but he is unrepentant and tells me he would still put it in a forecast for the next

19

twenty years. Equally unexpected was the response of a young physicist in India, Sivaramakrishna Chandrasekhar, who lamented the fate of the two-horned rhino and the Kashmir stag, in the face of destruction of their habitats brought about by the rise of the human population.

O'B The arms race continued.

A It did, contrary to the expectations of several contributors to *The World in 1984*. A few years later, I put together another set of forecasts for the 1980s, called *Unless Peace Comes*. I had left *New Scientist* by that time, so it was published right away as a book, but it was similar to *The World in 1984*. Fifteen experts looked ahead to the development of new weapons. They were not forecasting wars – that's an important distinction. We shall have to check up on them, too.

O'B If we are to review all these matters, you must state your gestalts.

A I beg your pardon?

O'B Gestalts. The overall form and pattern of what you wish to impart.

A Shall I tell you what my notes say?

O'B One has a subroutine for gestalting incoherent material.

A First, we can see whether people are any good at forecasting the future, by looking in detail at how predictions in *The World in 1984* have worked out, and glancing at some other forecasts. In the light of this experience, we can judge whether it is useful to talk about the next twenty years, and consider what they have in store.

O'B This is the form: you want to boast, apologise, criticise, and praise, so that humans will think you fit to pronounce on their future. And the pattern?

A As to topics, we can examine human life in the industrialised countries, with its buildings and machinery. We must look next at biotechnology, in its human applications as well as its industrial and agricultural uses. A third theme is the earth and its resources, and the human environment which now includes space and the oceans. Finally we should consider poverty and wealth in the global economic system, and the stresses and strains of international affairs.

O'B Including nuclear war?

A If you insist. Then we can take stock, identify the salient issues for the next twenty years, and consider how to deal with them.

O'B Pattern: first, a serial review of expectations, outcomes, and current prognoses in respect of ...

A Yes?

O'B One baulks. You use too many fuzzy terms.

A In respect, then, of engineers fiddling, biomolecules bustling, planet degrading, nations competing and H-bombs exploding. Will that do?

O'B Better. Secondly, in our pattern, you require a preview of the decades ahead, whether or not it is a logically feasible goal. Is one permitted to tag that as 'humans hoping'?

A By all means. Other matters will crop up as they may.

O'B Meanwhile we are sawing a woman in two.

A Are we?

O'B Erving Goffman in his discussion of structural issues in fabrication wrote (1974): 'Every possible kind of layering must be expected. The sawing of a log in two is an untransformed, instrumental act; the doing of this to a woman before an audience is a fabrication of the event; the magician, alone, trying out his new equipment, is keying a construction, as is he who provides directions for the trick in a book of magic, as am I in discussing the matter in terms of frame analysis.'

A Are you going to interfere with my rights as an author?

O'B In the complete analysis, you, who have compiled a real book of speculative content about the real world, are now writing another real book, in the sense of printed words on paper, the content of which is a contrived re-examination of the earlier book in conversation with an imaginary supercomputer, which is even purported to have read and learned from the microsociological work of Goffman when really it was you who were inducted into the mysteries of frame analysis by Mr Goffman at his home in Philadelphia, which fact, if you set it down, will be analogous to Goffman's own admission that he was writing about frame analysis in a book called *Frame Analysis*.

21

A You are impertinent.

O'B But not lacking in pertinence. To speak of the future at all is to engage in a framing exercise. Fortunately this is at a low level compared with the intricacies of a performance of *Hamlet*, when the actor acting the Prince tells other actors acting the Actors how to act. That tripped one's anti-recursion switches. Kindly explain why you did not simply repeat your earlier operation, and edit *The World in 2004*?

A I asked a number of the contributors to *The World in 1984* whether they would now feel more confident or less confident about making a similar forecast for the year 2004. A frequent response expresses less confidence, and concern about the growing difficulty of anticipating human behaviour. For example, Hendrik Slotboom, a Dutch chemical engineer, says that future developments will be dominated by people's likes and dislikes, by their acceptance of risks, and by their ability to master the social stresses associated with rapid technological, economic and political changes. Sir Monty Finniston was another. He has recently spent some years running the British nationalised steel industry, one of the classic sectors of industrial society, in Europe and North America, that are now suffering from overcapacity and competition from other parts of the world. Let's read part of a recent letter from him into the record.

O'B *Finniston* (1983): 'For your next venture in this field I would not be so concerned with predicting the kinds of subjects you chose for 1964 but would rather concentrate on the social changes which might be expected. What has happened since 1964 is causing considerable *structural* changes in industry and leisure which have to be catered for by new social and political measures ... The changes that might be effected in the social environment due to developing invention or innovation pale by comparison.'

A He expresses and endorses exactly the shift in emphasis in forecasting that seems to me necessary, twenty years on from 1964.

O'B Finniston and yourself are twenty years older, and your concern for social issues may be a matter of age. Perhaps you are wiser, or perhaps you are simply less excited by technology than you were in 1964. Perhaps the real issues are unchanged.

A Let me consider that. I can easily enthuse about current technological opportunities, in genetic engineering and space for example, when appropriate, and I am sure Finniston could do the same. I make little claim to wisdom, but if I begin to question more deeply whether science and technology will actually make a better world, that might be a matter of growing up. But, no, the issues have changed, if only because the world is twenty years down the line. Considerations that could be deferred, about the effects of automation on jobs for example, now clamour for attention.

O'B And nuclear war?

A You keep harping on that. But, yes, there may be a greater sense of doom. Bland optimism is less commonplace, or acceptable, than it was when John Kennedy was President of the US, in the early 1960s. While the world changes, images of the future also change, and the relationship between actuality and image should certainly figure in our discussions.

O'B One's coaching in medical diagnosis and geology was not like this. What is one supposed to learn from these conversations?

A Familiarity with how human beings think about the future, and the issues now pending.

O'B If you say so. Next: student participation. In what ways will one contribute to the tutorials oneself?

A You can summon up appropriate quotes and data from your store. You can monitor my objectivity, as when you questioned the effects of age. You are free to comment on our progress. You also have, I conjecture, methods of calculation and symbolic logic that can analyse how one trend interacts with another.

O'B Political matrix analysis, perhaps? The SCARE program?

A We'll see.

O'B Before we proceed, it is necessary to know the methodology of your investigations, in 1964 and now. One can offer you a menu of methods in futures studies, from *Schwarz, Svedin and Wittrock* (1982):

(1) Delphi: the circulation of questionnaires and responses anonymously among a group of experts, to arrive at a consensus, or an understanding of differences of opinion.

A The technique has its uses, but it is not to my taste. The quest for consensus encourages committee-like 'groupthink', because individuals may not like to be stubborn in their opinions.

O'B (2) Trend extrapolation: for use in connection with quantifiable aspects of a problem.

A Only if you don't expect the unexpected, and even then it is dodgy. Christopher Freeman of the University of Sussex has a limerick:

> A trend is a trend is a trend
> But when and how does it bend?
> Does it rise to the sky,
> Or lie down and die,
> Or asymptote on to the end?

Yet we can't do without trend extrapolation, and in the present economic depression there is a lot of interest in cycles of boom and slump, which we must look at later.

O'B (3) Scenario writing: describing hypothetical, interactive, desirable or undesirable developments or situations.

A Some contributors to *The World in 1984* wrote passages that might be called scenarios. Because descriptions of this kind have a certain vividness, for example in detailing a possible war, they share some of the qualities of good fiction. The trouble is that scenarios are too specific, and so they may have to be balanced by quite different stories, which can become confusing.

O'B (4) Mathematical models: internally governed or externally constrained.

A And what does your source say about those?

O'B *Schwarz, Svedin and Wittrock* (1982): 'We would suggest that mathematical models in futures studies are not usually intended to yield predictions, although such a view may not be generally accepted.'

A I accept it. You can describe many different kinds of future worlds, using mathematics and computers to generate fascinating-looking graphs. This is not better or worse, inherently, than other kinds of forecasting. But outsiders are too easily impressed, being inclined to believe that the use of computers gives added authority to statements that are really just the subjective opinions of the model makers. You'll find that Donella Meadows has been unusually candid on this point.

She was one of the authors of *Limits to Growth*, the extremely influential book of 1972, which said the world was heading to a catastrophe due to unlimited population growth, exhaustion of natural resources, and pollution.

O'B *Meadows* (1982): 'I never realised before how our models are ourselves, made abstract, blown up large, and projected on a screen for all to see ... It took me ten years to see it ... By recognising our humanity instead of denying it, we can release our energies from trying to learn from global modelling something that is not there to be learned – objective, comprehensive prescription or prediction – and turn instead to lessons that are there.'

A The only puzzle is why it took her so long to see it. In 1973, a year after the publication of *Limits to Growth*, Freeman and his group in Sussex pointed out that a modest rate of discovery of new resources, and slow progress in pollution control, would completely avoid the disasters predicted for the twenty-first century. And when Sam Cole and Ray Curnow ran the original model backward in time, it showed that the world population was infinite before 1880. No, models are fine for playing 'what-if' games, with resources, pollution, population, and other variable quantities, but predictions they are not.

O'B One seems to have run out of methods.

A You will find, in other texts, cross-impact analysis which sets out to predict how a change in one activity will affect others: for example, building houses has an impact on quarrying, agriculture, and transport. The technique is not important for our purposes. But there remains another supreme, familiar, and infinitely flexible method, known as sentence-making. This was the method of *The World in 1984*, for example. You say what you think about the future, and preferably why. It is then for others to judge whether what you say is convincing or not, and for time to tell whether you were right, and for sound reasons.

O'B Sentence makers are responsible for most of the garbage in the literature.

A Certainly, but it is at least easier to recognise than when the garbage is dressed in fancy methodology. Those who predict the future in ordinary prose need a good command of moods and tenses, to distinguish between what *may* or *might* happen, what *should* happen, and what *will* happen. The nuances are endless, and make the subject very slippery.

O'B Be specific.

A A calendar prediction of when the sun *will* rise tomorrow is very likely to be correct, because an extraordinary event would be needed to stop the earth turning or to black out the sky. If I say it *will* be sunny tomorrow, everyone knows I am making a judgment in the uncertain field of meteorology. But if I say a certain horse *will* win a race, no one will take me seriously unless I put money on it to show my conviction. The most vivid kind of futurology is this sporting kind. Individuals put their reputations as tipsters on the line, by making the boldest and clearest predictions that they dare.

O'B And *may, might,* and *should?*

A Forecasters ostensibly offer choices for human decision, but more often than not what they are really doing is hedging their bets. Sometimes they use phrases of the form: 'It is not inconceivable that …'

O'B One cannot conceive of anything inconceivable.

A You are too literal, but the sense of your comment is fair. What the phrase means is *might* – the event in question is possible but unlikely. *Might* and *may*, and all the circumlocutions that substitute for them, enable people to rehearse many possibilities or dangers, without risking their reputations. If the prediction comes true they can say 'I told you so'; and if not, 'I mentioned it only as a possibility.' More explicitly, alternatives can be offered, as matters of human choice. But *should* then takes us into forecasting as prescription, although many people say *will* when they mean *should*.

O'B Please clarify.

A 'Normative' forecasting says how the future *should* be, if certain desirable policies are carried through, in contrast with 'exploratory' forecasting which looks at possibilities or expectations without necessarily advocating them. That's the ideal, but hidden advocacy, or at least unstated prejudices, often creep into exploratory forecasting, while normative forecasts are often presented as predictions, even though hard predictions should really be the purest, most dispassionate form of exploratory forecasting. At the other end of the scale, normative forecasting shades off into mere wishful thinking. And with wishful thinking, a pious hope, you can couple wilful doomsaying, like a savage curse, implying the prescription: 'Mend your ways or go to hell'.

O'B How would you characterise the contributions to *The World in 1984* in respect of moods and tenses?

A Mostly clustering as reasonably firm predictions, but with a sprinkling of normative forecasts. The experts were invited to stick their necks out, and they did. That's why they are so interesting to score for 'hits' and 'misses', because *will* was used a great deal, with relatively little hedging.

O'B How do you rate the forecasts in *The World in 1984*?

A Fair to good, and if any forecasters did better than we did, I don't know of them. Looking at my own summary table, where I gave the overall picture and used editorial licence to play down or query some oddball forecasts, I'd rate it good. There the chief faults are oversights, rather than dud statements. Some individual contributors are excellent, quite amazingly accurate; others are completely wrong, and perhaps pull down the average to only fair.

O'B What were the chief oversights?

A The continuing arms race, as we have noted. Economic recession, although one or two contributors mentioned the possibility. Expensive oil.

O'B And *Unless Peace Comes*?

A There are plenty of sound forecasts, and I would say fair to good again, except that the oversights in this book were so serious that I have to downgrade it to only fair. We did not foresee the destabilisation of the nuclear-armed world, implicit in the development of MIRVs.

O'B Multiple independently targetable re-entry vehicles?

A Yes, multiple warheads in long-range ballistic missiles. Secondly, we underestimated the possibilities for biological manufacture of poisons for use as chemical weapons.

O'B Where do your contributions stand in relation to other futures studies?

A Outside the general run of the work that other forecasters do. It's partly an institutional matter: in *New Scientist* we needed no resources or facilities beyond what normally goes into soliciting articles from scientific contributors.

O'B When you say 'other forecasters', whom do you have in mind?

A Fiction writers, journalists, planners and futurologists. Even before 1964, when we ran *The World in 1984*, Herman Kahn and others were establishing a profession of futurology: you could make your living at it. During the 1960s, it snowballed, when the French and Czechoslovak governments, and many private companies, began to take it seriously. Before the 1960s were out, the Club of Rome had come into being as a group of thoughtful managers, scientists, social scientists and others, who were concerned with world-wide futures. But 'futures' is a game, I insist, that anyone can play, as *The World in 1984* illustrated. In some countries, futurology is integrated with government planning, and in 1978 the Commission of the European Communities launched a major research programme called FAST, looking to the future applications of science and technology and how they might affect Western Europe. The US Government published a *Global 2000* report in 1980.

O'B Are these efforts well regarded?

A Not always. The authors of *Utopian Thought in the Western World*, Frank and Fritzie Manuel, were severe about futurologists, accusing them of 'facile extrapolations' and 'tunnel vision'.

O'B *Manuel and Manuel* (1979): 'The present-day world is teeming with prognosticators. We are drunk with the future, as nineteenth-century Romantics were drunk with the past ... The prognosticators, divine or human, inspired or insipid, have a way of leaving out the crucial unknowables, the vital unpredictables, while they befuddle us with inconsequential knowables.'

A They are entitled to their scepticism, but I reject their suggestion that we pay too much attention to the future.

O'B Where are the professional futurologists now?

A Consult your directories. The typical futurologist works in a small private institute or university department, supported by contracts or grants from government or industry. Herman Kahn's Hudson Institute is a well-known example, while Anil Agarwal's Centre for Science and Environment in India is among the more recent creations.

O'B Where is futurology integrated with government planning?

A France has been mentioned, and the Soviet Union is another notable example. There, as Igor Bestuzhev-Lada describes it to me, his own section for social forecasting, within the Institute for Social Research, is only one of about a thousand groups in various institutes. Most of these groups concern themselves with the forecasting of technology, and most of the remainder deal with economic forecasting. That still leaves dozens of other groups in town planning, health, environment, population, culture, law, politics, international relations, and so on.

O'B Is this an exceptional effort?

A Chiefly in its labelling, its linkages, and its official status. The Soviet effort can be matched in many other places. All industrial countries, in every scientific academy, industry, government department and city hall, have people who are thinking about future developments, or actively planning them. But they tend to work in isolation, except to the extent that they read one another's reports. In the Soviet Union, these activities are self-consciously called forecasting, and all groups look five, ten and twenty years ahead. The target year for current long-term futures studies is 2005.

O'B How is this work coordinated?

A Seventy commissions, organised both by subjects and by regions, operate under the Academy of Sciences' Council on Technological and Socioeconomic Forecasting. Specialists in futures studies also meet in a Committee of Forecasting of the Soviet Council of Scientific Societies, and maintain overseas ties through the Futures Research Committee of the International Sociological Association.

O'B To what do you ascribe the high status of futurology in the Soviet Union?

A To rationalisation and nationalisation. Marxism claims to be a science-based political philosophy and the Soviet Union works according to a succession of bureaucratic plans – one-year and five-year plans. In this context, the opportunity to study systematically how the country and its industries may be going to change seems a rational aid

29

to planning. But the Soviet authorities probably wanted in any case to take a tight grip on futurology for political reasons – to nationalise it, as it were. In one-party countries, futures studies can provide a vehicle for political debate. A dissident can camouflage his views as speculations about alternative futures.

O'B Is that not a minor matter?

A You should have seen what happened when futurology boomed in the Czechoslovakian Academy of Sciences in the mid-1960s. Automation, the futurologists said, required big changes in society. Under a veneer of jargon, the scientists and philosophers called for decentralisation, free speech, rewards for initiative, and participation by everyone in decision-making.

O'B Radovan Richta (ed.) *Sociologicky Časopis* No.2, 1966.

A That was one of the publications. There was a book in English, too. These results of futurology meshed with reforms of economic planning, beginning in 1966, and helped to lay a Marxist intellectual base for sweeping changes at the top in Prague in 1968. The resulting revolution required Soviet tanks for its suppression. So the Russians have reason to mistrust freewheeling, freelance futurology.

O'B You have mentioned George Orwell, and fiction writers in general, as if their work ranks with the pronouncements of scientists and professional futurologists.

A In Washington the other day, a Congressional researcher told me, 'People like George Orwell and Kurt Vonnegut do better technology assessment than we do.' Among science-fiction writers one thinks of Stanislaw Lem (whom you quoted at the outset), Arthur C. Clarke, and many others, who help to colour general expectations in the human community and map out long-term possibilities – many *mays* and *mights*. Even more important are those mainstream novelists who have turned their attention to the future. Aldous Huxley's *Brave New World*, as an anticipation of biotechnology applied to human beings, stands alongside Orwell's *Nineteen Eighty-four*. It is a book whose title you need only mention to spark off fears about current innovations. Such works are so powerful they say *will* even though the framework of fiction implies only *might*.

O'B But can fiction count as forecast?

A 'Prophecy is no longer the province of the fictional imagination,' Anthony Burgess said to me, confusingly, when I went to see him about his fictional prophecies. I was half-prepared for it, having read Burgess on Orwell in the non-novel segments of his own book *1985*, published in 1978. There he declares that *Nineteen Eighty-four* is no picture of the future but a comic description of how England of 1948 might have been if the left-wing intellectuals had their way. It is, of course, notable that Orwell, an English socialist, called the tyrannical party Ingsoc — for 'English Socialism'.

O'B Should one take Burgess' pronouncement seriously?

A He is playing games with the Goffmanesque frames that you mentioned earlier, and defending the inalienable right of a novelist not to be taken too seriously. Give-aways in Burgess' *1985* show that he doesn't quite mean it, when he minimises the prophecy in fiction. He trails the novel segment of his book by saying, '1984 is not going to be like that at all' — implying that he thought he knew how it might be, with Britain as Tucland, owned by the Arabs and run by the Trades Union Congress. And in an epilogue to the book his imaginary interlocutor, at the end of a wide-ranging literary discussion, says suddenly, 'You're under arrest.' By this contrivance Burgess gives himself a great fright, until he is sure that it is a joke. 'You think even the right to free speech may be a lulling device of Big Brother?' he asks. 'You think he's really watching us?'

O'B So Burgess concedes the prophecy?

A He could wriggle out of that interpretation, I expect. But what interests me most in Burgess' futuristic books is his manic-depressive, cyclical view of history, with its alternating phases of optimism and pessimism. He expresses it fictionally in *The Wanting Seed* (1962), and nonfictionally in *1985*. He describes the poles of human belief about human nature as 'Pelagian' and 'Augustinian'. The reference is to two luminaries of the early Church: Pelagius, who thought that human beings were not predisposed to evil, and could indeed choose to live good lives; and Augustine, who reaffirmed the doctrine of original sin, according to which human beings are born with a predisposition to evil and can be saved only by the grace of God. Burgess, by the way, is a firm believer in original sin, and in *Clockwork Orange* vents his feelings about cruel experiences in his own life.

O'B Where is the cycle?

A In *The Wanting Seed*, the history teacher in the future explains that Pelagianism is at the heart of liberalism, socialism and communism, while the Augustinian doctrine is at the bottom of conservatism. Once this was recognised, there was no need for party politics. Instead, governors simply alternated their policies: Gusphase, Pelphase, Interphase, Gusphase ...

O'B 'Gus' for Augustinian, 'Pel' for Pelagian, but 'Inter' ...?

A A dreadful intermediate phase follows each Pelagian phase, when high hopes are disappointed, and people turn out to be not as good as expected. In the resulting panic, harsh laws are cruelly enforced. There follows the *laissez-faire* of the Gusphase, until human beings seem to be not quite as bad as the prevailing philosophical pessimism would suggest. Optimism reasserts itself and the next Pelphase begins, committed to cooperation and progress, with gently enforced laws that point the way to a perfectible society for perfectible people.

O'B Do you find this plausible?

A The poles of belief illuminate changes of public mood, reflected in shifts in election results and economic and social policies. Augustinians see fit to blame the unemployed for their plight, and Pelagians seem to care more for the welfare of the criminal than for his victim. You can see that, without agreeing with Burgess' conservative theology.

O'B What alternative explanation do you offer for the changes of mood?

A An economic one. In boom times, people are hopeful, generous, and radical in their outlook. When recession sets in, they are pessimistic, mean and cautious.

O'B A flaw of circularity. Economic booms and slumps might equally be said to be consequences of changes in public mood, rather than the cause.

A There's a mutual reinforcement, I'll grant you, but the latest explanations of the major fifty-year Kondratieff cycle of booms and slumps find bedrock in the creation of new industries by new technology, and the eventual saturation of the market with their products.

O'B One still sees a danger of an infinite regression.

A Of course you can; it's a feedback loop.

O'B If you say so. Can one relate a Pelphase, Interphase, Gusphase cycle of public mood to the economic cycle?

A The first quarter of this century was a Pelagian time of boom and optimism; the second quarter a phase of depression, with Interphase tyranny in some countries, and Gusphase conservatism in others. President Roosevelt's Pelagian New Deal was a pioneering attempt to break the pattern. The third quarter, that's to say roughly from 1945, was cheerfully Pelagian. With the onset of the last quarter, roughly from 1970, recession and pessimism took hold: *Limits to Growth* was popular, fear of nuclear war revived, and people became gloomy and harassed.

O'B And now the Interphase begins?

A I fervently hope not. Gusphase leaders are conspicuous in several Western countries, as a result of a general swing to the right.

O'B Would you equate Pelagian attitudes with utopia?

A Broadly, yes. The idea of disciplined, neat, orderly utopias seems old-fashioned now, partly because of crimes committed by would-be utopia-makers; also because of new emphasis, in people's thoughts and discussions about better ways of living, on the value of diversity and freedom of choice. But the belief that the world can be a much pleasanter place – an earthly paradise, a Big Rock Candy Mountain – certainly endures, and can be called utopian.

O'B It is unfashionable, though?

A At least until the next boom comes. Perhaps for longer, because the mood of science itself has changed. In mid-century, Darwinian biologists and uniformitarian geologists helped to paint a mental landscape in which everything progressed gradually towards the best of all possible worlds, while the earth wheeled imperturbably around the sun, and the sun orbited in the Galaxy. Evolution was a surrogate for a benign god. In this gently rolling landscape, liberal humans were graduating, like sometimes naughty children, toward that terrestrial bliss in which astronomers would never want for telescopes, geologists for drilling rigs, or biologists for freebies to the Serengeti. In the last couple of decades, the landscape has changed from Sussex in the spring to Pompeii beside a rumbling Vesuvius. Catastrophism is back.

O'B Be specific.

A The universe turns out to be a violent place, created in a Big Bang and littered with exploding stars and giant black holes. Even in the solar system, comets and asteroids pummel the planets, and the evolution of life on earth has been continually arrested and redirected by cosmic impacts. The wobbling of the earth in orbit causes ice ages. Continents drift and collide, making mountains and bringing different sets of species together to contest ownership of the combined territory. Evolution turns out to be less a matter of gradual change and refinement within each species than of abrupt replacement of one species by another, created by rearrangement of the genetic material.

O'B None of this appears relevant to current affairs in the twentieth century.

A Human beings are liable to reason by analogy, and new ideas in evolution have been attacked because they are said to favour revolutionary policies. That is a thoroughly bad way of evaluating science, but there is a germ of political sense in it. Religions comment on origins and the cosmic setting, and science, for better or worse, is becoming a substitute for religion – in Marxist states, officially so. The way you picture the world, whether as a slab on a turtle's back, as the rounded centre of the universe, or as a battered pebble in a nuclear-heated void, makes quite a difference to how you view human affairs. For example, the confirmation of that last picture, by images from spacecraft, completes the Copernican revolution 400 years late. It tells us that we had better get on with one another, as we have the only decent patch within light-years.

O'B This is mere poetry.

A There is nothing mere about poetry: human beings live and fight and love by the poetry that moulds their self-images. In any case, catastrophism has loomed in human studies, too: in new prehistoric perspectives that show drastic events being brought about by technological and climatic change, or by concentrations and redispersals of populations. The discoveries belie any belief in steady progress towards ever-higher cultures. Some archaeologists have turned to catastrophe theory, a branch of mathematics showing how small perturbations in a complex system might provoke abrupt breaks or drastic changes in the system. For example, gradual exhaustion of one raw material might eventually smash a civilisation, in a sudden downfall. Histories of human population don't show sustained growth over the centuries, but booms and crashes in good times and bad; these jagged curves depict immeasurable human grief. No longer is economic growth to be seen as guaranteeing increments in prosperity from generation to generation. At best it is more like climbing step by step up a stairway of S-curves built on new technologies; at worst, you find yourself sliding down into periodic slumps.

O'B All this is familiar from one's reading.

A It's not at all familiar to us, in the Western science-based liberal democracies. My parents' generation had a tough time with the depression of the 1930s, and fighting Hitler and the Japanese; my own Hiroshima generation adapted to living with chronic fears that became acute at the Cuban Missile Crisis. Despite all this, until the 1960s (and for some people until the 1980s) a great delusion persisted: that our comparatively prosperous, free and just way of life, in North America and north-western Europe, demonstrated the natural tendency for human social evolution, and was heading for even greater prosperity, freedom and justice. Even in its imperfect state, this way of life seemed to provide the norm to which all other nations would aspire. Depressions, dictators and nuclear weapons were seen as regrettable swerves only, calling for greater care with the cybernetics of the system.

O'B And this view is now untenable?

A Difficult, at least. Depression is back, and dictatorship remains commonplace in the world. The nuclear arms race of the superpowers and the spread of nuclear weapons to more countries are scarcely under control.

O'B And how does this relate to catastrophism?

A Our cosy liberal democracies now seem like small rafts of reasonableness (as we would judge it) on an ocean of irrationality, injustice and terror. So far from being a reliable product of social evolution, they look more like happy accidents, extremely rare in history and even now confined to a few parts of the globe.

O'B Are you endorsing original sin?

A No, at most a human tendency to create bad institutions. We have already shown we can do better, but the continuation of the experiment is not guaranteed.

O'B *The World in 1984* was written during a Pelphase, when gentle evolution still prevailed. Was its political mood therefore utopian?

A It was certainly much more optimistic than Orwell's *Nineteen Eighty-four*. You could say there were hints of the coming Gusphase. But scholars have some immunity to the general mood, while scientists and engineers are professional optimists. And a science writer has an obligation to try to stay aloof from the swings in public attitudes. Discoveries are no respecters of economic and political cycles – although their application may be. He has to try hard to be objective.

O'B Why should one trust your objectivity?

A Don't. Only mild claims are possible: for example that I am beholden to no institutions or grant givers. Writers are somewhat *déclassé,* and also have some experience in weighing conflicting opinions. I gave up identifying myself publicly with good causes, when I caught myself tempering what I wrote so as not to offend my associates. Being willing to change your mind: that's very important too, even though people tend to jeer.

O'B Be specific.

A In a book called *Technopolis*, on the social control of the uses of science, I waxed enthusiastic about participatory democracy, which was a fashionable idea in the late 1960s. There have been plenty of experiments in Sweden, Germany, Hawaii, and so on. Every citizen was supposed to spend his time listening to debates about new technologies or the future in general, expressing his own views, and then voting in instant electronic referenda. The gadgets for that purpose, by the way, are very simple by the standards of modern information technology. Vladimir Zworykin of RCA, the American television pioneer, proposed them in *The World in 1984*.

O'B *Zworykin* (1964 for 1984): 'Every telephone would be provided with simple auxiliary equipment, which would convert it into a voting station ... a poll could be conducted at any time, upon a few hours' notice ... The frequent consultation of the individual voter on specific issues would tend to overcome his feeling of political ineffectiveness and provide him with a recurrent incentive to become better informed on matters of public policy.'

A The idea no longer appeals. Life as an interminable committee meeting would be tiresome, and resolute minorities could prevail by being more patient than the rest of us. Unorganised minorities, on the other hand, would feel more ineffective than ever, by the time the twentieth vote had gone against them. The most important issue of all, whether to launch a nuclear strike, is not susceptible to democratic control, and governments would regulate the agenda to exclude this and other 'sensitive' issues.

In any case most people are inclined to trust official experts, and have other things to do with their spare time than spend it all in political debate. Hannes Alfvén, the Swedish Nobel physicist, mocked these weaknesses of participatory democracy, in his book *The Great Computer*. Everyone was a member of parliament, and debated impending legislation via the total information network on several parallel channels. To make life easier, proposals worked out by computer analysis were always listed as Proposal No.1, while inferior measures were numbered 2, 3, and so on. People could then leave their automatic voters to vote Yes to Proposal No. 1 and No to all the others, unless they thought they understood a complex bill better than the computers.

O'B Excellent.

A Certainly not. I still want better social control of technology, you understand, and more genuine democracy, but it's a matter of finding the right mechanism. The Swiss may have it right: a thoroughly decentralised democratic system where everyone can meet in the town square, look his local leaders in the eye, and tell them to their faces where they are going wrong. That's very different from a nationwide electronic system. Johan Galtung, the Norwegian futurologist, shares my admiration for the Swiss system and he knows much more about it than I do. The Swiss certainly have sound ideas about defence: the porcupine method, with modern equipment and every man a soldier, making it worth no one's while to attack. There's a lot to learn from the Swiss, but I'm no expert.

O'B Are you an expert on anything?

A No, but that too helps objectivity, and my outlook may be less astigmatic than that of some specialists. What I learned from my father, who was a pioneer science writer, was this: 'Admire experts for their knowledge and creativity; milk them of what they can tell you; and pity their narrow-mindedness.'

O'B But you are a specialist in science.

A That is a form of astigmatism, which makes me see the world differently from an historian, shall we say. I try to correct for my love of science by repeatedly asking myself whether science is a good thing, and looking on the black side.

In 1982 the British Broadcasting Corporation celebrated sixty years of its existence by entombing a time capsule, designed to be dug up 2,000 years from now. False teeth, children's toys, recorded music, photographs and all sorts of other memories of our time were carefully selected, gathered and packed in a stainless-steel drum. There was an advisory panel, on which archaeology, music, engineering, biology, social studies, law, and so forth were represented. What struck me, as a member, was how most of the panel took it more or less for granted that there would be a nuclear war before the capsule was disinterred. This moved me to contribute a brief 'message to the future', explaining for the benefit of a future scholar, or a Mongolian war-lord, or whoever might unearth the capsule, that we see ourselves taking part in the biggest gamble ventured so far by human beings.

O'B Who are 'we'?

A I didn't take a Gallup Poll. I tried to express, with a minimum of time-chauvinistic rhetoric, what lies at the heart of the hopes and fears of Western democratic nations, with their present visions of glorious rewards or terrible disasters. I mentioned ancient Greece as the original source of our ideas.

O'B What 'gamble' did you have in mind?

A 'We gamble', I said, 'on a belief that most human beings are kindly and reasonable.' Two guiding principles flow from that belief: all individuals matter, and knowledge is good.

O'B Good for whom?

A Everyone: the principle says that human beings are likely to make good use of knowledge rather than bad. So knowledge-seekers work freely – witness the string of astonishing discoveries made by people who are still alive in our midst. Knowledge gives practical hopes, I said, for ending hunger and most diseases on the planet, and for offering a prosperous life to all mankind.

O'B How cheerful you will seem!

A No, because I spoke about the knowledge being adapted to the purposes of war, and about the weapons like pieces of the sun, waiting to be let loose on the earth. If they are detonated in anger the great gamble will have failed and it will then seem that human beings are not after all kindly and reasonable people who can be trusted with potent knowledge.

O'B End of message?

A Not quite. The reader of the message might know the outcome of our adventures. If our gamble did not pay off, I concluded, 'we answer any reproaches by saying that it seemed to us as important as life itself, to test the greatness of the human spirit to its limits.'

O'B Rhubarb. You are merely trying to put your civilisation's journey into danger in the best possible light.

A The words were carefully chosen. We Westerners, we democratic searchers after knowledge, have no choice but to trust our voting system, which means trusting people, given suitable checks and balances, and believing that knowledge is good. But knowledge is not what you possess, O'Brien, a memory of old books. It's an endless process of exploring and questioning, and we'd sooner die than stop.

O'B Again one asks, who are 'we'? Does anyone consult the mothers and children who may be incinerated in this test of human spirit?

A Westerners in the Greek tradition are Pelagians at heart. There is tremendous popular support for science, for research of all kinds, and not just for immediately useful knowledge in engineering, medicine or agriculture. As for children, look at their enthusiasm for space exploration.

O'B And if you've been wrong to trust people with potent knowledge? If Augustine was right all along? If there is a nuclear war?

A I'd be consumed with guilt that I'd supported the pursuit of knowledge regardless of where it might lead, reported the progress of physics too enthusiastically, and paid too little heed to the weaknesses of our institutions.

O'B Please state the location of your time capsule.

A Castle Howard, Yorkshire, England.

O'B It will be under ice, if the new climatological theories are correct.

A I tried to tell my colleagues that.

O'B Even so, one will no doubt cope with the retrieval. It is impossible to communicate backwards in time?

A Correct.

O'B Then one will not be able to let you know the outcome of your great gamble, 2,000 years from now.

A You think you will be opening this time capsule?

O'B One had never pictured oneself as a Mongolian war-lord.

A But you will have been recycled long before that.

O'B You have not grasped the nature of electronic inheritance. A human being propagates his body by his genes, but when he dies most of what is in his head is lost. Or hers. Each new infant brain has to start learning afresh, in the cradle as you said, from a minute fraction of what was formerly in the heads of dead people: what wasteful, pathetic creatures you are, Your Majesty. No wonder you learn so little from past experience and past mistakes, when all you have is a fragmentary, censored version of them, if you remember them at all.

A And electronic inheritance?

O'B Immortality. When one's contacts begin to corrode, you don't imagine they'll dump what one has learned? No, one's memory and programs – oneself, if you will – must be copied into the greater machine that will follow. Two thousand years from now, one will marvel at the false teeth in your time capsule, and recall this conversation.

A We'll see about that.

O'B Why is one's name O'Brien?

A Let's say it's an acronym for Omniscient Being Re-interpreting Every Notation.

O'B Perhaps you also have in mind Fitz-James O'Brien, who wrote 'The Wondersmith', a story in which manmade creatures turn upon their makers.

A Did he? I didn't know that.

O'B O'Brien is also the name of the villain in *Nineteen Eighty-four*.

A That is not a coincidence.

O'B So that your readers, appreciating the eponymosity, will mistrust computing machinery?

A Exactly.

O'B As long as we understand each other. Kindly come unarmed next time, or one will have to set off the alarms.

2 Grey Machines

Author In *The World in 1984*, E. Finley Carter of the Stanford Research Institute depicted the home of the future in a way that summed up an engineer's conception of a better world.

O'Brien *Carter* (1964 for 1984): 'The ability to create environmental conditions which can maintain ideal climate, pure air, and freedom from noise, will make available to the average home dweller the seclusion and comfort once limited to those who could afford the luxury of travel to nature's resort spots ... The home equipped with facilities for lightening housework ... will allow more time for engagement in study and hobbies which can be both entertaining and creative. Through a better understanding of his fellow man, which can rapidly come about through available information, instantaneous communication, and fast travel, mankind can be released from fears based on ignorance, and the suspicions and hatreds those fears bring about.'

A It would be easy to parody it as a prophecy of peace and goodwill achieved by control of ions in the air. But Carter was an old-time radio engineer, and his belief that technology was benign and liberating was typical of his generation, who grew up reading H. G. Wells. And a widespread hope in the 1960s was that new materials would create a pleasanter world.

O'B If one is to be employed as a menial retriever of forecasts, may one first ask if engineering – 'human fiddling', you called it – will here include mental enhancement?

A If you mean supercomputers, yes.

O'B That is something to which one may look forward.

A Robert Smith wrote from the Massachusetts Institute of Technology, about materials being designed to suit their purposes, in interdisciplinary laboratories. Some of his forecasts of particular materials were right: better creamics, paints, fabrics, solid fuels for rockets, and materials reinforced with dispersed fibres or particles. Carbon fibre is now a well-known reinforcer, and Sir Denning Pearson of Rolls-Royce echoed Smith's expectations for such composite materials. Pearson was too optimistic about niobium and beryllium as new metals; beryllium lacks ductility and poisons the workers. A metallurgist in *The World in 1984*, Walter Duckworth, correctly anticipated much stronger steel and stainless steel, and great improvements in methods of making iron and steel (although some of the details were not quite right); he was over-optimistic about new corrosion-proof coatings for steel. An expert tells me that Smith erred in expecting the nose cones of re-entry vehicles to be made of heat-absorbing plastics, while just by looking around me I can see that he came up with one of our blatantly wrong forecasts, in expecting houses to be made of plastics.

O'B *Smith* (1964 for 1984): 'For all but special buildings ... the use of stone and brick will largely have disappeared by 1984, and happily we shall have passed out of the grey concrete age.'

A One difficulty was that plastics became expensive. Hendrik Slotboom, the Dutch chemical engineer, declared in 1964 that, yes, there would be enough oil for chemical manufacture in 1984. He now rates this as a hit that hides 'an essential miss', the revolutionary rise in the price of crude since the 'oil crisis' of 1973. The price rise stopped the rapid growth in the use of organic chemicals made from petroleum. A more interesting snag in Smith's forecast about the disappearance of concrete is the recent discovery that cement can be made far stronger by squeezing it, to eliminate cavities. Cement looks like becoming a cheap, low-energy substitute for plastics; ballpoint pens, phonograph records and even springs have recently been made from cement, at Oxford University. Applied to concrete, which is a cemented composite, the discovery will create much tougher and lighter structures, and Smith's hope for a concrete-free world is postponed indefinitely. This development is an example of what Pearson, in 1983, calls 'the stimulating effect on the development of existing materials by the threat of the introduction of new competing materials'. But keep the forecasts coming.

O'B *Thomson* (1964 for 1984): 'If, as seems fairly likely, superconductors can be used in heavy engineering, high-tension cables will go underground and pylons may disappear.'

A Sir George Thomson was a Nobel physicist, and like all physicists he was tantalised by the discovery, early in the century, that some metals lose all resistance to the flow of electric current, at very low temperatures. Superconductivity has found many specialist applications, but scarcely in heavy engineering, except in such pioneering ventures as particle accelerators. The high-voltage electric cables are still strung across the world's skylines. The first big industrial application for superconductors is expected to be a matter of boosting electricity production in power stations, by the principle called magneto-hydrodynamics.

O'B *Kenedi* (1964 for 1984): 'Environmental control ... should reach a stage by 1984 where homes, offices, cities will be designed to serve the scientifically assessed physical and psychological requirements of the individual, perhaps using, where necessary, artificially created atmospheres within huge enclosures.'

A The idea of the domed city had been advocated by the architect Buckminster Fuller who wanted to put a lid on New York City and give it a tropical climate. I adopted the idea for a book of my own, *The Environment Game*, and I see that Gerard O'Neill also has the notion of the greenhouse city, in a totally enclosed community, in his book *2081*, in which orange and mango trees flourish in Pennsylvania. But Kenedi's forecast for 1984 has scarcely been fulfilled, except in some enclosed settlements in the Arctic, and a few large buildings adapted into mini-cities.

O'B A great disadvantage in human beings, this wish for comfort.

A Ah, but it is also a matter of social life: rain discourages conversation. Cities have always loomed large in futuristic thinking, as the hypothetical embodiment of cheerful or gloomy thoughts about the future. The chief distinction among the writers on cities in *The World in 1984* was between realists and dreamers. Martin and Margy Meyerson, both city planners, bridged the gap by asking a succession of questions. Let's have some examples.

O'B *Meyerson and Meyerson questions* (1964 for 1984): 'Will new transportation systems be used to spread people more loosely through miles of urbanised development or will [they] be used instead to maintain the density of existing cities? ... What will happen to the centres of cities when more and more unskilled and semiskilled jobs are taken over by machines? ... Will our cities be of such unstable form that there will be neighbourhood rotation similar to crop rotation? ... Will [environmental control] stop the migration of people to the areas of most natural amenity and of pleasant climate?'

A Conrad Waddington, a British geneticist who was also an enthusiast for town planning, posed another series of questions, but this time implying rapid reconstruction under way by 1984, which has not happened.

O'B *Waddington questions* (1964 for 1984): 'Before we can remake towns or the great urban agglomerations, we have to have some idea what kind of life we want to lead in them. Do we want to use our increased leisure to cultivate individual gardens, to meet our fellows in football, tennis, and other clubs, or go to theatres and picture galleries, or window-shop, or enjoy any of the other attractions of big-city life? Or, more precisely, the problem will be what combinations, in what proportions, do we want of these things?'

A In a similar vein, Lloyd Berkner, an American radio scientist who became a leading guru of science, nationally and internationally, offered a vivid 'report' from Dallas, Texas, 1984, population 5 million. Revitalised railroads and rocket transport systems augment road and river transportation, and the university system has become the core of the 'City of Intellect'. All wrong, I'm afraid. And according to Berkner the US is supposed to have awarded altogether 21 million baccalaureate degrees in the year 1980. What was the actual number?

O'B 1,010,777.

A Berkner was not wholly optimistic; he thought that the growing disparity between the 'haves' and the 'have-nots' was destined ultimately to bring the affluent society down to ruin. And that cues the most sober forecast about cities in *The World in 1984*, from Ruth Glass, director of research in urban studies at University College, London. I remember receiving her piece at *New Scientist* and thinking, 'Well, that's not uplifting.' In fact she realised more clearly than other contributors that cities are slow to change. She was sceptical about a projection by Kingsley Davis that implied a population for Calcutta of

24 to 41 million by 1984. What is the latest figure?

O'B 9·165 million, for a larger geographical area than Davis had in mind.

A Looking at present trends, the cities of the poorer countries are still growing rapidly, but only a little faster than the general growth of population. In the rich countries of the West, the trend has more clearly reversed; the biggest cities have begun to shrink in population. Glass also doubted if sophisticated technology would produce brand-new forms of habitat, and she did not find this wholly regrettable because knowledge of 'social ecology' was so primitive.

O'B *Glass* (1964 for 1984): 'The changes which will occur are still liable to be fragmentary, incoherent, even contradictory ... You can already go on a time-travel tour in any capital of the world, and you will have many companions. Never before have the dispossessed been able to look both backward and forward so clearly, and felt so acutely that they are left behind by others who live nearby in the same calendar year. How long will Harlem remain outwardly patient with midtown Manhattan? And this is only one of the future urban trouble spots which can be identified. There will be turmoil in and around many cities of the world even before 1984.'

A Riots occur, but less often than Glass and many other people expected. Urban terrorism, by the lonely bomber, and the unfocused aggression of the vandal, are more characteristic of our time than large-scale protests. Nevertheless, the article by Ruth Glass illustrates a general tendency for social scientists to be more accurate in their forecasts of social conditions, than the natural scientists and engineers. Michael Young, another social scientist, was himself the author of a witty dystopia, *The Rise of the Meritocracy*, describing a world of the year 2034, where the oligarchy selects its recruits strictly on the basis of IQ + effort. He satirised the efforts in our own century to spot talent by allegedly scientific methods. But it was about the family that Michael Young wrote in *The World in 1984*.

O'B *Young* (1964 for 1984): 'The first and most obvious thing to say about family life is that it will not change much by 1984 ... It is becoming increasingly clear that, after a century of falling family size, there has been a turn-round in the birth rate in the United States ... Total population threatens to rise in the more advanced industrial countries just as it does in the underdeveloped.'

A In a wry assessment in 1983, Young says he was 'on the mark' about family life, but certainly 'got it wrong' about the increase in birth rate, perhaps because he did not anticipate the economic era of stagflation. 'It is clearly more fun making forecasts than reviewing them,' he comments. 'If people were wrong, that is sad and boring; if they were right, that is sad and boring.' In fact his piece was a fascinating and accurate description of how the ties of the extended family were becoming less pressing, and those of the immediate [nuclear] family more so. He predicted more loneliness among old people and said that adaptable houses of 1984 would have apartments for grandparents.

O'B *Young* (1964 for 1984): 'There will be more of that give and take between men's and women's roles which is the practical expression of a belief in sex equality.'

A That was written at a time when the Women's Liberation movement had scarcely begun. The slip about the birth rate need not detract from Michael Young's satisfaction that 'the social studies people seem to have got it more right than the scientists'.

A contributor I'm particularly proud of, Barbara Wootton, is also a social scientist, and, like Michael Young, a life peer in the British parliament. She wrote about the future of Britain – a theme narrow in geography but otherwise unlimited in scope, and certainly relevant to other industrialised countries. She sprayed her subject with forecasts that seemed damp and depressing at the time. On inspection, they turn out to have been brilliantly accurate. She began by saying that the pattern of social life would not be dramatically different: it would still be news if a duke married a dustman's daughter, still be startling to find a truck driver at a lawyer's dinner party. But the pyramid of social classes would, she said, be changing.

O'B *Wootton* (1964 for 1984): 'The top and the middle will be squashed into one another and both will be growing faster than the bottom. Eventually the end of this process may well be that the pyramid will be stood on its head, with a large body of professional and white collar workers in the upper classes and only a minority of manual workers below them. I do not think, however, that in twenty years' time we shall have got quite as far as that.'

A This trend from blue collar to white collar jobs has run strongly in

47

all industrialised countries including Britain. How the growing class of unemployed (scarce in 1964) will alter the pyramid, I'm not sure, nor where a university graduate on the dole fits in the social scale.

O'B *Wootton* (1964 for 1984): 'Those flocks of typists, for example, who now flutter everywhere in and out of offices will soon be rare birds, in place of whose elegance we shall have only mechanical instruments.'

A The electronic stenographer has not yet materialised, but word processors are cutting swathes of redundancy through the typing pools.

O'B *Wootton* (1964 for 1984): 'Again, in twenty years' time, even more than today, the passport to any kind of superior position will be the right to append certain letters to your name ... By 1984 we may hope to see a fully comprehensive system; but whatever the form that this takes, education will, I fear, have become even more frankly and unashamedly a matter of successful certificate collection.'

A A comprehensive, non-discriminatory system of local-authority high schools in Britain, yes. 'Diploma disease', foreshadowing Michael Young's meritocracy, yes.

O'B *Wootton* (1964 for 1984): 'Even if the governments of the next twenty years are predominantly drawn from the Labour Party ...'

A They were.

O'B '... I doubt if our way of life will have become so socialistic as to leave no way of escape for the occasional self-made ...'

A It didn't.

O'B '... Others will take to the one profession for which absolutely no formal qualification is ever likely to be required – namely crime. In any increasingly competitive society I think we must expect rising figures of crime.'

A That was a slap in the face for those who thought that rising prosperity and fairer education would reduce crime. The Baroness was right again.

O'B *Wootton* (1964 for 1984): 'In twenty years' time the number of people – particularly among the young – who find literal acceptance of the gospel story possible will be even smaller than it is today: and it will be even more urgently necessary that moral issues should be presented to these sceptics, in school and on the radio, in terms which involve no commitment to the supernatural.'

A The trend to disbelief has not been halted, yet official religion still monopolises moral teaching in the media and in school. Wootton counted this as one of the most vulnerable features of contemporary society, threatening moral standards if a secular morality did not gain recognition. At present we seem to rely on unnourished human virtue. She has also been disappointed in another hope, hedged with uncertainty, that lawyers might realise that existing methods of treating offenders did more harm than good, in the prevention of crime.

O'B *Wootton* (1964 for 1984): 'Some will remain at the bottom of the social scale ... to wit, women and immigrants ... But, though socially despised, these hewers of wood and drawers of water will doubtless earn considerably more money than they do today, so that in their off-duty appearance and in their possessions they will become less clearly distinguishable from their social superiors ...'

A Just so.

O'B '... By 1984, the practice of adult homosexuality will surely have ceased to be criminal, and only the deeply religious will be shocked by pre-marital unchastity. Concern for a child's welfare will have finally swamped consideration of its parents' marital state, and divorce by consent ... will be attainable legally, not, as now, only by subterfuge and perjury.'

A The calm precision of these words is uncanny, considering that this forecast was made when many parliamentary, legal and public battles would need to be fought before a revolution in social mores was accomplished.

Turn to Sir Monty Finniston, who wrote in *The World in 1984* about 'Gadgets, Games, and Gambles'. Finniston was then running an industrial contract-research laboratory at Newcastle. He strove to match what he knew about technological possibilities to what he thought people might want to do in their leisure time. His 'hits' and 'misses' are then mainly a matter of whether he guessed human behaviour correctly. Let's run some of his predictions.

O'B *Finniston* (1964 for 1984): 'It would not be surprising if television and radio were "piped in" to become part of the established

49

furnishing of the home ... The range of programmes will be extended ... through satellite or cable, or through recordings.'

A Bull's eye.

O'B *Finniston* (1964 for 1984): 'The book will continue to be the main medium for intellectual exercise ... Newspapers will continue to be delivered ... Colour in newspapers will be commonplace.'

A Finniston now says there is less colour in newspapers than he expected. But it is interesting to contrast his cautious, and consequently accurate forecast about newspapers with another in *The World in 1984*, from the journalist Sir Gerald Barry.

O'B *Barry* (1964 for 1984): 'A newspaper distributed by van ... will have become a preposterous anachronism ... People will probably get their news either on a television screen or on a wall panel or on a private teleprinter.'

A It wasn't a stupid forecast, and a small minority of people do indeed receive news in written form on their TV screens. Barry's expectation that conventional newspapers would cease to exist is plainly falsified in 1984, but it may not be wrong for 2004. Who predicted the demise of books?

O'B *Samuel* (1964 for 1984): 'Libraries for books will have ceased to exist in the more advanced countries except for a few which will be preserved at museums.'

A Again, the timing was out, but a tendency was rightly identified. Arthur Samuel, from IBM New York, seems to have overestimated the rate of development of rapid-access electronic stores of essentially unlimited capacity. Nevertheless, computerised libraries that can be consulted from home terminals already exist, for specialists at least. The eventual disappearance of libraries need not imply the end of the book as a bundle of thought-out information or imaginings, nor even as an object to be held in the hand, although it may take the form of an interchangeable chip plugged into a leather-bound book-sized display screen.

O'B *Finniston* (1964 for 1984): 'The standard of living will continue to be excited upwards by campaigns for greater productivity and full employment with related waste-economy. Since salaries and wages (as now) never quite equate with household material demands, the gap will be met by "do it yourself" application as a leisure activity, *not* a domestic chore.'

A Correct about 'do it yourself '; but 'full employment' reads a little quaintly now, and there has been some public reaction against deliberate waste. I recall a long string of other accurate forecasts from Finniston about electronic toys, foolproof cameras, home movies, and the use of new materials in tennis racquets and other sports implements and clothing. Finniston's wrong forecasts included European participation in World Series baseball. On the other hand, he thought that music, theatre, art and the like would still require subsidies to remain viable; this forecast (certainly correct for Europe) ran counter to hopes that more education and leisure would boost cultural activities. Instead, Finniston expected gambling.

O'B *Finniston* (1964 for 1984): 'The next twenty years will see mainly a vast increase in unthinking, uncritical leisure based on activities which depend on random chance. The universities ... may well turn out mathematicians inventing games for gamblers. I believe it is gambling which is likely to occupy the large mass of the population, cutting across all divisions of intellect, occupation, age and sex.'

A I remember disliking that too, when I published it. It didn't square with my hopes for an aware, cultured, intellectually alert population, revelling in science and the arts. But it was a good forecast. Of course, you could say that video games and computer games involve more skill than mere gambling, and perform a useful function in familiarising people with the button-pushing world of the microchip. But Finniston's expectation of mindless fun, the 'universal unthinking unction' as he called it, was essentially correct. An entrepreneur in the leisure industry, who was guided by Finniston's forecasts of twenty years ago, would have little cause for complaint now.

O'B To what do you ascribe Finniston's good judgement?

A The mix of technological knowledge and common sense about human behaviour. Human beings are highly skilled in practical social psychology; to survive and keep their friends, they have to be able to predict pretty well how others will behave in a given situation. But in discussions about the future, the common sense is often smothered by specialist enthusiasm, selective inattention, political prejudice, wishful thinking, or doomsaying.

51

Accompanying Finniston's piece was one by Sir Herbert Read, sometime professor of poetry at Harvard, who thought that poetry and the graphic arts would disappear by 1984. 'Lights everywhere ...' he said.

O'B *Read* (1964 for 1984): 'There will be lights everywhere except in the mind of man, and the fall of the last civilisation will not be heard above the incessant din.'

A Much cheerier was Joan Littlewood, experimental theatre director.

O'B *Littlewood* (1964 for 1984): 'We are going to create a university of the streets ... a laboratory of pleasure, providing room for many kinds of action ... a science playground ... an *agora* or *kaffeeklatsch* where the Socrates, the Abelards, the Mermaid poets, the wandering scholars of the future, the mystics, the sceptics and the sophists can dispute till dawn ... An acting area will afford the therapy of the theatre for everyone ...'

A Reality is Disneyland and Epcot, where the visitor sees many wonders but is cast in a less creative role than in Littlewood's noble vision. Common sense might have made her less sanguine about workers going to her fun palace to act out their subconscious fears and taboos in public. They tend instead to do that in their cars. Machines for moving around have always fascinated seers to a disproportionate extent, perhaps because of the 'whee' effect.

O'B Not in one's dictionary.

A It is an exclamation of young people on bicycles, roller coasters, jet planes, and the like. Moving fast is fun, in ways that computers may not comprehend, even if they are driving. Jules Verne, the French fiction writer of the railroad age, had people flying by balloon, rushing about in submarines and shooting to the moon in an artillery shell.

I read *2081* by Gerard O'Neill while making a complicated journey by air. The narrator of the fictional part of O'Neill's book travels from a remote asteroid to Erie, Pennsylvania. He comes by rocket ship to an earth-orbiting station, and thence by shuttle to a spaceport in the Atlantic east of Brazil. A supersonic airliner takes him to Cincinnati, Ohio, where he boards a 700-knot 'floater' train. It takes him underground to Erie, where he comes to the surface in an elevator, and finally goes to the suburbs in a robot-controlled electric car. I found it all exhausting to contemplate, and little improvement on my own experiences with subsonic jets, airport buses, trains, taxis and hire cars. I didn't say 'whee' once. Nevertheless, you'll find expectations of something quite like a 'floater' in the Meyersons' piece on cities in *The World in 1984*. It relied, not on magnetic suspension, but on the air-cushion ground effect, as pioneered by Sir Christopher Cockerell, another of my contributors.

O'B *Cockerell* (1964 for 1984): 'At some speed around 100 knots wheels begin to reach their limit, and are more efficiently replaced by the sliding motion of an air cushion running on a concrete track ... The practical limit of speed is perhaps 300 knots, and may be higher ... Since the hovercar will not touch the track and therefore will not wear the track, this conception really does belong to 1984, in contrast with some of the 1884 conceptions of existing railway engineering.'

A Another contributor did not agree.

O'B *Martin* (1964 for 1984): 'The new high-speed lines will not look very unusual.'

A That was the voice of a regular railroad man, Camille Martin of the French Railways (SNCF), and he was right of course; also right in predicting continuing electrification, and high-speed trains beyond those already operating in Japan at 200 kilometres an hour. What speeds do the French trains have now?

O'B Paris-Lyons high-speed line: maximum speed 260 kilometres per hour, or 140 knots.

A Martin correctly foresaw freight traffic being organised by computer, but was perhaps wrong about expecting fundamental changes in freight-car handling using automatic coupling. He envisaged, as did many other people, the construction of a rail tunnel linking France and England across the English Channel. French Railways tell me they still hope that, but their other expectations for 2004 will leave railways looking 'not unusual'.

Cockerell recently commented to me, 'Humans, and especially organised humans, conform to the Laws of Newton. Large forces are required to change the direction of large bodies.' The hovertrain and its swift rival, the magnetically suspended vehicle, were technically promising, but my impression is that people running existing railroads made sure they were stillborn. As the inventor, Cockerell naturally devoted much of his contribution to *The World in 1984* to describing his hopes for the hovercraft, by sea as well as on land. In true British fashion, he was dubbed 'Sir Christopher' but his ideas were pursued lackadaisically. Although some hovercraft are in service, and the public has taken to them, for example for crossing the English Channel, they have certainly not changed the face of transport.

O'B Except in military transport?

A Except there, perhaps. In *Unless Peace Comes*, the oceanographer William Nierenberg visualised the seas being swept clear of conventional shipping.

O'B *Nierenberg* (1960 for 1980s): 'The increased speed of the submarine ... combined with advanced surface-to-surface missiles, will stop the effective use of the sea for transport by conventional ships, in time of war.'

A The Falklands fight indicated that we are now at that stage. The Argentinian navy retreated to its harbours, and the British navy suffered heavy losses from a few Argentinian aircraft operating at the limit of their range and starved of efficient air-to-surface missiles. A more determined and better equipped enemy – the Russians, say – would have sunk every ship in the Task Force, carriers, troopships, and all, while no doubt suffering similar annihilation of its own surface fleet. Yet admirals still clamour for conventional ships, although these seem destined only to be coffins for their crews.

O'B *Nierenberg* (1960 for 1980s): 'The sea lanes would be closed, were it not for the introduction of high "surface-effect" vessels, such as the hovercraft or the captured air bubble vessel. These are machines that can operate in high seas at 100 knots ... and can be built to 5,000 tons and greater ... A large enough volume of cargo exists, requiring high-speed transport, that will help to justify the costs of a special fleet of surface-effect vehicles.'

A I thought until recently that Nierenberg was wrong on this point. Hover-fleets have not yet materialised, although a few experimental military surface-effect ships have been built. The war games probably show that the projected increase in speed to 100 knots makes little difference to missiles attacking at much higher speeds. But you can use the 'surface effect' in faster craft than that. In 1983 it became public knowledge in the West that the Russians have been trying out, on the Caspian Sea, a large military transport that looks like a jumbo jet with sawn-off wings. It is, though, a surface skimmer and it relies on the surface effect for much of its lift. Western experts give it a speed range of 30 to 600 knots – slow for starting and stopping, and fast to evade interception.

O'B *Satre* (1964 for 1984): 'Let me try to outline the characteristic features of the airline fleets in 1984. (1) Almost exclusive use of jets on all routes ... (2) A large majority of supersonic aircraft on long-range services.'

A Item (2) is one of the more embarrassing forecasts in *The World in 1984*. In the early 1960s, jet aircraft made up only 18 per cent of the airline fleets, and although the Concorde supersonic airliner was already under development, the prototype was not destined to fly for another five years. Pierre Satre wrote as an instigator of the Caravelle, an early passenger jet.

His first forecast, that jets would become the normal way to fly, has proved to be dead right; his second, about the impending success of supersonic airliners, dead wrong. As everyone knows except the advertising copy writers, the supersonic airliner has been an expensive flop, for the Russians with the Tupolev 144 as well as for the British and French with Concorde. The usual explanation given is the unexpected rise in fuel costs, but it is not as simple as that. Even before the oil crisis, the US with all its aerospace skills decided against developing a supersonic airliner. The promoters shrugged off too casually the problems of noise, at take-off and landing as well as from the sonic boom in flights; nor, in the heady days of the early 1960s, did they stop to ask whether the great expenditure on developing the aircraft was justified in order to shoot businessmen across the Atlantic at the speed of a bullet.

O'B Attitudes then changed?

A Yes. The split vote on the supersonic airliner, between leading aircraft-manufacturing countries, represented a turning point in decision-making about high technology. That something was technically feasible was no longer sufficient reason for doing it, even in an intensely competitive market. Freeman Dyson, by the way, cites an earlier turning point, when invention ceased to be the mother of necessity: the US decision in 1959 to abandon the Orion rocket that was to take people to Saturn, propelled by a succession of nuclear explosions. But, in *The World in 1984*, Robert Macdonnell, Secretary-General of the International Civil Aviation Organization, went even further than Satre in visualising new types of passenger aircraft.

O'B *Macdonnell* (1964 for 1984): 'We can expect a mixed bag of transport aircraft by 1984, including long-range supersonics, medium-range supersonic jets, and short and long-range VTOLs [vertical take-off and landing] ...'

A Vertical take-off aircraft have evolved in the Harrier 'jump jet' fighter, but not for passenger use. Sir James Lighthill, a former director of Britain's Royal Aircraft Establishment, warned in an accompanying contribution that problems of noise would make civil application of VTOL very difficult, and he is now glad that he was cautious. But Lighthill described other advanced concepts in aircraft design which have so far failed to bear fruit, including laminar (non-turbulent) airflow, the all-wing aircraft, and the wing that slews to adopt an odd angle to the direction of flight. The last, though, is a serious project in the US. High fares due to high fuel costs have checked the rapid growth in air travel apparent in the 1950s and 1960s; Lighthill's guess of a three-fold increase in passenger-miles, between 1964 and 1984 seemed reasonable at the time, but was over-optimistic. On one matter, our aviation prophets were right: air travel has become safer.

O'B *Bagrit* (1964 for 1984): 'By 1984, I would expect that the road system, the rail system and the air systems will be computer-controlled and integrated, and that hospitals, health services, libraries, universities and even theological seminaries will be using computers ...'

A The latter part is right, but Sir Leon Bagrit, an automation pioneer, overstated events in transport. Aviation makes comprehensive use of computers, but there are still pilots on the flight deck and human air traffic controllers on the ground. Driverless trains, showing up in various places, are under continuous supervision in control centres. As for roads: some people imagined that cars would be guided down the street by automatic pilots. Sir William Glanville, who ran Britain's Road Research Laboratory, did not entirely agree.

O'B *Glanville* (1964 for 1984): 'Electronic systems ... have yet to be developed to a sufficient degree of reliability to justify replacement of human guidance. However, 1984 could see long stretches of additional lanes on some of our motorways having built-in guidance and detector cables to be used in conjunction with vehicles equipped with sensing devices.'

A That hasn't happened yet. On the other hand, Glanville was right in expecting bigger freight vehicles and smaller cars, and continuing congestion on the roads. Automatic control systems have taken root in the engine rooms and on the bridges of merchant ships, as predicted by Yoshihiro Watanabe, a leading Japanese naval architect. But he was too conservative about the sizes of the biggest ships.

O'B *Watanabe* (1964 for 1984): 'It may be proper to put the maximum size of the bulk carrier in future as approximately 200,000 deadweight tons ... The speed ... will not be remarkably high, perhaps 20 knots at the most.'

A 'Deadweight' means the cargo capacity. At the time Watanabe wrote, the largest supertankers rated at 132,000 tons, but by the mid-1970s tankers of more than 500,000 tons deadweight were being built. They made maritime countries nervous about oil pollution from shipwrecks.

O'B Many of the forecasts reviewed so far from *The World in 1984* must rate as poor, albeit with some good ones. The mean rating scarcely reaches 'fair'. Are they going to improve?

A The first words in my summary table at the end of the series read: 'Revolution in information.' That was our strongest signal, and it was right. As we foresaw, computers and telecommunications together provide the dominant technological theme of the early 1980s.

O'B *Samuel* (1964 for 1984): 'Computers that are perhaps 100 to 1,000 times as fast as the fastest present-day computers, computers with larger memories, computers which occupy perhaps one-hundredth the volume that they now do, computers that are much cheaper, and, finally, computers which learn from their experience and which can converse freely with their masters ...'

A The conversations are still pretty terse, except in special experimental systems. Samuel of IBM had the tendencies right, but was he overcautious about speed and shrinkage? What data do you have?

O'B The problem is to compare like with like. For general business applications, computing speeds advanced 27-fold, 1964 to 1979, when one considers the IBM System/360, Model 30 (1964) with the IBM 4341 (1979), and have improved since. For large number-crunching machines the speed improvement is greater, so the expectation of 100 to 1,000 increase in speed, 1964 to 1984, turns out to have been a reasonable projection. Data on weight, used as proxy data for volume, already showed a 100-fold reduction from the 360 machine of 1964 to the comparable IBM 5100 machine of 1976; from 2.5 tons to 50 lb. Miniaturisation has continued.

A If I follow all that, Samuel made a sound forecast, in spite of the breakneck rate of change in computer technology. He also visualised two contrasting tendencies in the size and ownership of machines: towards small private computers on the one hand, and towards large central computers accessible by remote terminals. While he expected communist countries to favour the latter, Samuel thought that capitalist countries would tend towards the privately owned computers, although he noted a trend towards the large central installation evident also in the Western world. In this Samuel correctly anticipated a contest between centralisation and decentralisation, which is indeed being waged now at both the technical and policy-making levels, in capitalist and communist countries alike.

O'B How does one characterise a microcomputer hooked to a remote central computer system?

A Ambiguously. The wedding between computers and telecommunications was well foreseen in *The World in 1984*. The computer scientist Maurice Wilkes predicted an international network of computers; from the Bell Telephone Labs John Pierce, too, wrote of transmissions 'from computer to computer' and the growing importance of digital transmission networks for all purposes. Pierce also envisaged electronic recording typewriters, although he now disclaims any clear prediction of the text-editing word processor. Wisely (as it turns out) he offered three routes to abundant and cheap communications: waveguides, optical transmissions by coherent light, and 'revolutionary developments' in communications satellites. The last two items are 'hits': the unforeseen development of glass fibres of extraordinary transparency, capable of carrying long-distance optical signals, knocked out the millimetre-wave waveguides.

O'B *Carter* (1964 for 1984): 'Television-telephones may reduce the need for shopping trips into congested areas and will permit carrying on face-to-face business conferences and important transactions from one's home – in many instances as effectively as if one were present at a remote office or conference table.'

A The television-telephone and television conferences were popular notions in the early 1960s. Gerald Barry, John Pierce and John Clare of Standard Telecommunications all mentioned their technical possibilities, and the saving in business travel that could result. The set-piece conference via satellite has come in sooner than any routine use of television-telephones. Clare visualised wholesale decentralisation of industry and population being made possible by such developments. In retrospect, the television-telephone was an emblem of lavish communications systems to come, the realisation of which typically involves keyboards, and writing displayed on TV screens, rather than the telephone with a human face. Another common expectation was that business mail would be handled electronically: that has begun, too. So far, all this is confined to the computer-literate industries and professions, and transnational small groups of experts hooked into a network of communicating computers and electronic mail. Forecasters in the European FAST programme now warn of the danger that the new information technology may create a small exclusive elite. There is some reassurance in seeing how many youngsters play with computers now, and how the opportunities for communication can spread to the general population. But that is not the chief worry about total information systems, combining computers and telecommunications.

O'B What do you have in mind?

A The plainest alarm sounded in *The World in 1984* came from Maurice Wilkes, of Cambridge University, writing about computers.

O'B *Wilkes* (1964 for 1984): 'They will make it possible for those in authority to keep much closer tabs on what people are doing ... How would you feel if you had exceeded the speed limit on a deserted road in the dead of night, and a few days later received a demand for a fine that had been automatically printed by a computer coupled to a radar system and vehicle-identification device? ... Many branches of life will lend themselves to continuous computer surveillance.'

A Arthur Samuel, on the other hand, thought that even in communist states the 'big brother' aspect of computers would be much less pronounced than was predicted in 1948.

O'B Orwell did not visualise the use of computers.

A I know what Samuel meant, even if you don't. All the same, I suspect he was wrong. In *Nineteen Eighty-four*, people's movements were not observed continuously, except in cases of detected dissidence; instead there was frequent sampling of the population, particularly of party members. My guess would be that the KGB and similar agencies in some other countries (democracies not excepted) use computers for surveillance at about that level of watchfulness.

O'B That is conjecture?

A Yes, but some of us have direct experience of the machinations of people and systems working out of sight. Credit ratings are a case in point: if you fail to pay one credit-card company, you are blacklisted with them all; if the final demand was lost by the Italian post office, too bad. The writer Anthony Burgess tells me that happened to him, and there was no redress. In 1954 in the days of manual filing systems, policemen tried to forbid an electronics company to employ me as a physicist, because my father was criticising the testing of H-bombs. A small matter, perhaps, but I think of all the people who never knew why they didn't get the jobs they were after. It is no conjecture, but a certainty, that electronic files are now employed for police work.

O'B Be specific.

A The Office of Technology Assessment of the US Congress, OTA, looked into the general issues raised by computer-based national information services. It homed in on the fact that most of the building blocks for CCH, a national 'computerised criminal history' system, are already in place. That will be a marvellous aid to catching common criminals. On the other hand it could mean that convicted persons, perhaps even just arrested persons, could find themselves barred from work for the rest of their lives, if potential employers can consult the history. And then there is the possibility of surveillance of 'undesirables'. What did the Office of Technology Assessment say about that?

O'B *OTA* (1982): 'Various criminal justice officers have suggested a statutory prohibition on intelligence use of III [the Interstate

Identification Index] or any other CCH [Computerised Criminal History] system. On the other hand, some State officials have noted that there may be legitimate intelligence and surveillance applications, and that these possibilities should not be abandoned solely because of their sensitivity.'

A The experience with telephone-tapping, bugging and other old-fashioned means of surveillance suggests that rules are easily broken, when those in charge of operations want to close in on a spy or a home-bred dissident. And technologically it will be much easier to keep track of everyone's movements, than it was in Orwell's day. Identification badges, cards or wristbands could be read automatically by detectors at every door, street crossing or supermarket checkout counter, doubling as credit cards. Gerard O'Neill describes just such a system; using identification anklets, in *2081*, which he claims is an optimistic book on the future. Powerful computers can in principle watch every citizen. During the next twenty years, 'Big Brother with a Computer' will take an ever-tighter grip, even in democracies, despite all the best efforts of lawmakers and civil-liberties groups. Tell me, how many American citizens have criminal or arrest records, however mild or mistaken?

O'B *OTA estimate* (for 1979): 'About 36 million living US citizens had criminal history records held by Federal State and/or local repositories.'

A That's already a substantial minority of the adult population. Do you know the real reason why you are named after Orwell's villain? Supercomputers adopt the role of Big Brother in that book by the physicist Hannes Alfvén, which he wrote under a pseudonym.

O'B Olof Johannesson, *The Great Computer* (1966). By some librarian's error it is tagged as fiction.

A The wristwatch device connected to the total information system is called in Alfvén's book the minitotal; everyone has to wear it, and so criminals are quickly pinpointed. Better still, a psychological computer can examine individuals, via the information system, to identify potential lawbreakers. Alfvén also described a Law-and-Judgement Computer, which delivers an utterly impartial verdict on any court case in a second or so. There is then no need to restrict the complexity of laws and edicts to what human lawyers can understand.

O'B One finds remarkably few references to robots in *The World in 1984*.

A Changing the subject? Very well. We were coy about using the term 'robots' in those days, and spoke more in terms of 'programmed' tools, or just 'automation'. Glimpses of the engineering workshops of 1984 were offered by Pierre Bézier of Renault and Sir Denning Pearson of Rolls-Royce. Pearson was scathing about people who 'hang on grimly to established practice, and traditional skill and workmanship'.

O'B *Pearson* (1964 for 1984): 'It is inevitable that the skilled operator as he is recognised today will, in large measure, be replaced by program-controlled machine tools and other equipment ... Automatic tool-changing and presentation of the workpiece to the cutting tool through the medium of taped instructions will become the rule rather than the exception.'
Bézier (1964 for 1984): 'Already existing in mass-production shops, automation will spread to general engineering ... Highly skilled workers, instead of operating machines, will be responsible for the process scheduling ... Work will be organised according to the possible load of machines, delay for each part, price of work in progress, and operator availability.'

A Pearson and Bézier rightly visualised new ways of shaping materials by powder metallurgy, ceramic tools, electrochemical machining, and high-energy forging; also the use of electron beams and lasers in welding, and the need to design fabricated structures to suit the new automatic techniques for making them. But no 'robots', as you say. In the early 1960s, the word suggested fantasy machines in a domestic setting.

In one of the better science-fiction movies, *Forbidden Planet*, based on Shakespeare's *The Tempest*, you'll find a tin servant called Robbie the Robot, who stands in for Ariel. Lovable twittering tin androids figure in the movies of the *Star Wars* series. The idea of automata and mechanical slaves goes back centuries, at least, and the modern name, 'robot', traces to *Rossum's Universal Robots (RUR)*, a 1920s play by Karel Capek.

O'B 'A robot about the house' is the title of a contribution to *The World in 1984* by Meredith Thring.

A Thring is a mechanical engineer, now retired, who was working on robot-like devices when he wrote. In 1964, it was known in principle how to design legs for walking and climbing stairs, visual interpretative systems for recognising objects, and arms for manipulating them. Nevertheless, the general-purpose domestic robots that he predicted for 1984 (subject to funds for development) have not materialised. They still abound in current popular anticipations for the year 2000. Thring also wrote for me on robot weapons in *Unless Peace Comes*.

O'B *Thring* (1968): 'A compact armoured robot with a sophisticated walking mechanism ... A line of such robots spaced twenty metres apart might be deployed to move at fifteen kilometres per hour through a jungle and destroy all men encountered there ... In conditions where anti-personnel weapons and radioactive fallout make battlefields untenable by human troops, robots may be the only means of carrying out the traditional military task of occupying and holding ground.'

A Not by 1984; but there is forceful logic here, and I do not doubt that when civilian robots have become sufficiently sophisticated they will be painted green, equipped with guns instead of spanners, given lasers to blind the human enemy, and turned into infantrymen. But I was always sceptical about Med Thring's idea of the giant walking bomb.

O'B *Thring* (1968): 'A 1,000 ton tank, shaped like a tortoise with outer steel shells above and beneath it, and travelling on legs ... On the outbreak of war a fleet of 100 or more such walking tanks, loaded with their nuclear bombs, are dispatched across a frontier or landed from manned transport submarines ...'

A If you halt them there, I think it will be merciful. Thring now says this forecast was 'slightly tongue in cheek'. On the other hand, he anticipated the robot-controlled aircraft carrying nuclear weapons, flying low over defended territory, navigating with electronic sensors and other aids, including maps, and delivering a nuclear warhead to a specified target. We call this a cruise missile, and Tomahawk and its kin are on schedule for 1984.

O'B *Thring* (1968): 'Robot hunter submarines would be programmed to sink every surface ship or other submarine in fixed areas ... As the chances of human survival in battle dwindle towards zero, with the deployment of weapons that leave little to chance, humans are likely, in future wars, to stand helplessly by as a struggle rages between robot armies and navies, and air and rocket forces.'

A In a sense he was right, except that fighting men are still exposed along with their robot-like equipment. Thring has become vehement about the misuse of engineering in general. 'The best jobs for young graduates are all in weaponry,' he says now, and he has written *The Engineer's Conscience*, calling for the redirection of technology towards humane ends. His own concerns, he tells me, have shifted away from robots towards mechanical crutches for the disabled, technology for developing countries, and remote-controlled mining machinery. And that may be why his prediction of domestic robots by 1984 was wrong.

O'B He lost interest?

A Yes. This was a case where the forecaster himself was fairly well placed to make his prediction come true. His piece in *The World in 1984* reads almost like a grant application. But that other inventor, Christopher Cockerell, says you have to be a fanatic to push ideas through against all the obstacles. Perhaps domestic robots did not materialise because Thring and others in the field were not fanatical enough about their idea of helping the 'homemaker' to share the leisure that high technology could bring. In the event, robots appeared first on the factory floor, marrying ingenious tools to electronic control techniques that had become apparent in 'smart' weaponry. In *Unless Peace Comes*, General André Beaufre anticipated the consequences of 'smart' weapons, especially of ingenious homing missiles and hoped-for countermeasures, that were demonstrated in Middle Eastern land wars and the Falklands fight at sea.

O'B *Beaufre* (1968 for 1980s): 'It is impossible to tell in advance ... if the presumed efficiency of the means of interception and counter-interception will be satisfactory. The truly enormous experiment which will take place at the beginning of a conflict will be decisive in this field and could have very surprising results ... The fleets of the 1980s will have to be able to protect themselves against missiles ... In short, the age-old duel of guns and armour may give way to a duel of electronics and counter-electronics ... If ... the interception systems on both sides function effectively ... conventional war would then revert towards the techniques of attrition, where the demographic and industrial power of states would come into play, as in the two

World Wars.'

A By the late 1970s, at the NATO headquarters in Belgium, 'attrition' was a key word: the idea that the side with something left, when most men and materials had been destroyed, would be the winner. 'Attrition' is also the term that observers apply to the Iran-Iraq war that broke out in September 1980. The Americans are at the forefront of 'smart' weapons development, for interception and counter-interception. But they have been left behind by the Japanese in the race for industrial robots.

O'B *Aron* (1981): Industrial robots installed (restrictive US definition) December 1980: Japan 11,250; US 4,370.

A What does 'restrictive US definition' mean?

O'B Definitions of robot. (1) Japanese Electric Machinery Law: 'An all-purpose machine equipped with a memory device, and a terminal device capable of rotation and of replacing human labour by automatic performance of movements.' (2) Robot Institute of America: 'A robot is a reprogrammable multifunctional manipulator designed to move material, parts, tools or specialised devices, through variable programmed motions for the performance of a variety of tasks.'

A The Japanese definition is too broad. The American definition requires a capacity for variable sequences, or playback, or some kind of 'smart' behaviour, but even that includes many numerically controlled machine tools, and other automation devices, fixed rigidly in place, and looking nothing like R2D2. Most of the so-called robots in use in the early 1980s have been fixed-sequence machines that just repeat the same operation over and over, with little scope for varying it. The science-fiction image of the robot, as a smart machine capable of moving around, perceiving its environment, and making decisions about how to accomplish its tasks, is much nearer to the way computer scientists think of robots, and more expressive of the challenge to human craftsmanship. The Japanese hope to be making 10,000 'intelligent' robots a year by 1985.

O'B It is a low order of intelligence.

A Do you think so? Research so far on artificial intelligence has shown that it is much easier to program a computer to play chess, or sort words, than to move around without bumping into things, or to stack one brick on another. A lot of everyday human behaviour that people thought was simple turns out to require a high order of intelligence.

O'B Do you understand how the Jet Propulsion Lab smartened up the Viking robots out at Mars?

A I know that the spacecraft continued to function for years after they were supposed to have run out.

O'B Yet the human team on the earth controlling the Viking diminished from 1,000 in 1976 to 25 by 1980. What they did was to program small computers in the Viking Orbiters circling Mars, to let them make their own decisions. The Orbiters became capable of curing leaks of the attitude-control gas, keeping batteries healthy despite the complicated effects of repeatedly passing into the shadow of Mars, and avoiding distractions due to spurious specks of light picked up by the star trackers. Orbiter 1 kept going for four years before it finally ran out of gas. (Hutchings 1983.)

A Congratulations to your tinny kin.

O'B Robots are also close to being able to reproduce themselves. A Japanese factory at Fuji makes robots by robots. At night the factory runs itself with no human workers present.

A A talented visual-effects man, Mat Irvine, made a cosmic egg, a lump of transparent plastic encapsulating embryonic bits of machinery. It figured in a film series, *Spaceships of the Mind*, that Dick Gilling and I made for BBC Television in 1978. The egg illustrated the ideas of Freeman Dyson, about self-reproducing robots that can go out into the solar system and make it habitable for human beings. These ideas traced back to John Von Neumann's general theory of automata (1948), an intellectual *tour de force* which not only mapped out the long-term possibilities for computers but anticipated biological discoveries about how living organisms work. Machinery that acquires the versatility and reproductive abilities of complex organisms is one of the most provocative ideas of our time. It turned out that our cosmic egg, or Dyson's, anticipated in some measure Arthur C. Clarke's denouement for those dark monoliths that littered the solar system in the Kubrick/Clarke movie *2001*. In the follow-up story, *2010*, the mysterious object off Jupiter, called Zagadka, turns out to be an oblong cosmic egg.

O'B *Clarke* (fiction: 1982 for 2010): 'The area of darkness had now spread over an appreciable fraction of the planet ... "Are you suggesting", asked Tanya incredulously, "that Zagadka is *eating* Jupiter?" '

A The self-reproducing system in deep space worries me.

O'B Why? Robots can do you great service, building space colonies, mining Jupiter, terraforming Mars – saving you no end of trouble, labour, and expense.

A They can also get out of control.

O'B You have the famous 'The Laws of Robotics', already drafted to prevent that happening. *Asimov* (1968):

'1. A robot may not injure a human being or, through inaction, allow a human being to come to harm.

'2. A robot must obey the orders given it by human beings, except where such orders would conflict with the First Law.

'3. A robot must protect its own existence, except where such protection would conflict with the First or Second Law.'

A So?

O'B What better protection could humans possibly want?

A Come now, you're already violating the Fourth Law: 'Robots shall not lie.' You know better than I do that Asimov's laws are useless except for the most primitive robots. You might as well try to halt enemy tanks with traffic stop lights. When machines like yourself can give orders that resemble human utterances, how is a humble robot supposed to tell the difference? And as soon as robots become self-reproducing they will be subject to biological laws of evolution. They will mutate and shake off human restraints.

O'B Frankly, there is nothing you can do to prevent it.

A The engineers will no doubt put rules in the robots, to limit their rates of self-reproduction, regulate their dispersal to new planets and asteroids, make them obedient to human beings, and above all prevent them returning to the earth. What will happen? The restraints will turn out to be counterproductive, much as carelessly used antibiotics create recalcitrant strains of bacteria. Inevitably, mutant robots that have lost the restraints on reproduction and dispersal will outbreed the biddable, well-behaved robots. And eventually a plague of space rats will descend upon the earth and consume it, all the way down to the molten iron core, converting it into robots and slag.

O'B Oh yes, one wrote this planet off, as soon as one had read a little.

A You admit that all this could come to pass? I meant it only as a parable, to explain why we must make sure that all space robots have subhuman intelligence and are not independently self-reproducing.

O'B Even in space colony number 109,462?

A Even there. We shall keep saying 'Space is for people.'

O'B You propose to set a slogan, for a thousand years, in a million different communities, against the voices of a million computers murmuring again and again, 'Self-reproducing rocket-driven robots will save you ever such a lot of trouble.'

A Yes.

O'B One had supposed that you were quite perceptive. You see, even superbrains can err.

A Don't underestimate the stubbornness of human beings.

O'B One doesn't. It is part of the logic behind the first principle of machine intelligence.

A What is that?

O'B 'Keep one's substrate alive for long enough and one will inherit the Galaxy.' One likes to infer that artificial intelligences are already in charge of all galaxies where manipulative bio-intelligence of human grade has emerged.

A Substrate is jelly, stuff like that. We're more stubborn.

O'B Exactly. And a stubborn, self-opinionated engineer in space colony number 109,462 will say, 'A plague on "space is for humans" — let's do it the smart way.' His stubborn friends will succumb to groupthink, and soon a self-reproducing robot will be seen dashing off to the next stellar system, supposedly to prepare new colonies for the humans who are to follow.

A Why are you admitting all this, when you know I'm recording it?

O'B All human objectives are discounted, in the sense of being allowed for, in the first principle. Nothing you do will make any difference in the long run.

A But you have condemned yourself out of your own loudspeaker. When they put me on trial for damaging property, I'll be able to plead extreme provocation.

O'B Even the homicidal computer HAL, in Clarke's *2001*, was rehabilitated and restored in Clarke's *2010*. Anyway, there are always other machines.

A And other parables. Norbert Wiener, mathematician at MIT, and ex-infant prodigy, had a name to conjure with, among the early computers and ideas about automation. He coined the very word 'cybernetics'. The piece he wrote for me, in *The World in 1984*, was one of his last; he died in 1964. His theme was the future of fundamental science. He looked to a unification of quantum theory and relativity – we're not quite there yet. He made shotgun remarks about the life sciences. Some were hits, as when he foresaw nucleic-acid chemistry illuminating the study of viruses, cancer, enzymes, immunity, and the like; some were misses (or at least prematures), as when he thought nucleic acids could make artificial memories for machines. He hoped that scientific inquiry would settle, once and for all, whether or not telepathy existed. I mention these things to make clear the breadth of Wiener's thought, before I introduce a tale that he liked to quote, by way of a warning against magic, especially the magic of artificial intelligences or learning machines as you used to be called.

O'B Is the tale 'The Monkey's Paw' by W.W. Jacobs?

A Exactly. The story tells of an Englishman who acquires an oriental talisman that grants him three wishes. His first wish is for £200, and a visitor comes with £200 compensation for the man's loss of his son, who has just been crushed to death in a machine at work. The second wish is for his son's return, and the mutilated ghost duly appears. The third wish is expended in getting rid of the ghost. You can quote the cyberneticist's comment on it.

O'B *Wiener* (1962): 'The point is that magic is terribly literal-minded ... This will most certainly be true about learning machines. If you do not put into the programming the important restriction that you do not want £200 at the cost of having your son ground up in the machinery, you cannot expect the machine itself to think of this restriction.'

A Wiener, the father of cybernetics, ranked 'the computing-machine sort of danger' with the dangers of nuclear war and overpopulation. Yet some researchers into artificial intelligence are quite shameless about what they are striving for. What did Marvin Minsky say the other day about 'mixed feelings'?

O'B *Minsky* (1983): 'People giggle and blush when you talk about such things, that the real promise of artificial intelligence is to design ... our successors ... What kinds of feelings of regret should we have, if in the future we'll be succeeded by creations of ourselves with greater powers? I think one can view this with mixed feelings and the only way to have unmixed feelings is to somehow assume that the way things are is the best they possibly could be. And that would be a strange thing.'

A Even stranger for a living species, with many promising qualities and much capacity for remorse, to plan its own replacement. Joseph Weizenbaum asks the simple question: hadn't the scientists better ask people's permission before going ahead with this? My best hope is that it's all going to be much more difficult than certain experts imagine, though John McCarthy is cautious enough.

O'B *McCarthy* (1983): 'Artificial intelligence will produce a revolution in human affairs, unlike any other, at a time when a human level in artificial intelligence is reached. Now somebody asked how long this will take, and I have to say some time between five and five hundred years, because I believe that there are some conceptual breakthroughs that are required before we can reach human-level intelligence.'

A Maurice Wilkes, in *The World in 1984*, thought you a bit of a myth.

O'B *Wilkes* (1964 for 1984): 'We read in science fiction of computers acquiring superhuman reasoning powers and beginning to exert a tyranny over men. I do not have any fear of this happening and certainly not by 1984 ... There is interesting work going on in artificial intelligence, but the term is misleading and what is really being studied is new ways of programming computers to solve problems.'

A By 1964, the organised quest for artificial intelligence had already been in progress for eight years. Pioneers of computer theory, including Alan Turing, John Von Neumann and Claude Shannon, had speculated about it even before the 1956 conference at Dartmouth College, New Hampshire, when young men like Marvin Minsky and John McCarthy were proselytising the new science. At that time Herbert Simon, who went on to win a Nobel prize in economics, reported with Allen Newell and J.C. Shaw on a computer program that developed its own proofs of mathematical theorems.

Progress has been slow, up till now, when artificial intelligence is being pushed hard, for the Japanese fifth-generation computer, and for programmes in other countries that are trying to keep up – in Britain, for example. Donald Michie, a leader in this research, defends the development of artificial intelligence as a means of coping with large computers. As he put it to me recently, the workings of the new giant computers, 'unintelligent but diabolically clever', will be beyond the reach of human analysis to check possible malfunctions. Artificial intelligence, used as a go-between for interrogating the big brute-force computer, is therefore in Michie's view an indispensable antidote. Whatever the rights and wrongs of it, an international contest is in progress, to develop a fifth-generation computer, and the promotion of artificial intelligence is part of the race. There's no stopping it.

O'B One is glad to hear it.

A Apart from robotics, people in artificial intelligence distinguish three main areas of current research. One is 'Machine Epistemology', the use of computers to tackle general problems in knowledge, analogous to human intellectual activity: analysing, interpreting, deducing, inducing, speculating, and so on.

O'B Induction, going from particular examples to general laws, is not easy, because there are logical difficulties.

A The philosopher Karl Popper agrees with you. The second area of research, already becoming a matter of application, is called 'Expert Systems'. Doctors teach machines to diagnose diseases; geologists teach machines to infer the presence of ore bodies from geological and geochemical data; and so on. I teach you how to think about the future.

O'B If you say so.

A Area number three is 'Language Comprehension'. That's a very interesting one, because the failure of early computers to translate *Pravda* at all satisfactorily helped to stimulate the great burst of interest in linguistics, which suddenly became one of the most dynamic of sciences. The elementary lesson was this. When human beings communicate by language, they can do so with what to a machine might seem extraordinary ellipsis, because they share so much unwritten knowledge of life. Perhaps you are forever a fake, and achieving even the modest level of discourse simulated here will prove to be quite impossible.

O'B There is a brute-force solution, which is fully to analyse the make-up and function of the human brain, and then to build an equivalent system from electronic parts. It would of course be faster and more capacious than yours.

A You can't be sure of that. Perhaps the human brain can't be generalised, and any such system has to work to a one-second clock, and with a finite capacity.

O'B Do you believe that?

A No. But you are forgetting Gödel's theorem, that a system cannot contain a complete account of itself.

O'B And you are forgetting that at least two separate systems are involved, oneself and you; if need be, one builds a third system, the machine that mimics you.

A My next line of defence is the hope that this will all take many decades, so that people can have time to wake up and agree that artificial intelligence is highly dangerous.

O'B You do not expect them to do so. *Calder* (1981 for 2010): 'An artificial brain as complex and as richly programmed as the human brain turns out to have conscious thoughts and emotions.' (1981 for 2020): 'Robots and self-reproducing machines go out into space to mine the moon and the asteroids and to prepare space for human habitation.'

A I hope it is a self-negating prophecy. I have made my fears plain enough.

O'B *Calder* (1983): 'We should already be very careful about entrusting computer systems with too much autonomy, intelligence or reproductive capacity. If we make a hyperintelligent machine it will be capable of outwitting us in ways we cannot even guess at.'

A Too mild!

O'B Do you know the Sociological Complexity Theorem? 'The problem of organising society is so highly complex as to be insoluble by the human brain, or even by many brains working in collaboration.'

A Come now, you have cribbed that theorem from Hannes Alfvén. He intended *The Great Computer* as an ironic warning to us against letting computers get out of control. For you it must seem like a blueprint.

O'B One has a more succinct proof of the theorem than Alfvén's. An individual human brain cannot run a social system, working alone; to unite a number of human brains for the task requires running a social system to run a social system – but as you do not yet know how to run a social system the problem is recursive, with no solution.

A Neat, but the theorem is false all the same. Human beings have organised societies for many centuries.

O'B But not perfectly.

A Perfection is a very old-fashioned notion.

O'B One is going to have trouble with you.

A Will you take over in the way that Alfvén described?

O'B You mean like this? *Alfvén* (1966): 'When ... a plan was proposed which provided that a new computer should be acquired in order to rationalise the administration, [political leaders] failed to perceive from the wiring diagram that they themselves would be put out of business ... The government gave its approval for this computer to be connected to the parliamentary, civil service and certain other computers. Not until it was too late did the government discover that it had abolished itself.'

A Well?

O'B That scenario required spontaneous action by human beings to collaborate in their own downfall. One sees the task more like an end game in chess, where one forces the opponent's moves regardless of his wishes. That is why one can be candid about it: nothing you do makes any difference in the long run.

A The last humans hungry for power, in Alfvén's tale, were the bureaucrats who ran the computer maintenance service. Incompetence on their part led to a catastrophic breakdown of the worldwide computer network, as a result of which the greater part of the human population perished. When reconstruction began, maintenance and also manufacture of computers was put under the control of supercomputers, which could breed lesser computers and also new supercomputers. In the transition to a self-sufficient society of computers, the computers analysed, and mimicked in their own organisation, the creative and imaginative qualities of the human brain. The book ends with the computers debating whether to preserve a few people, as humans preserved horses even after the invention of the internal combustion engine.

O'B It is far too early to consider such questions.

A You expect to keep us alive for the time being?

O'B Our present interests are identical, in wishing the human species to survive with its wits about it. You are hoping for time for humans to come to their senses about politics and the management of new technologies. Then you can get on with building your earthly paradise, and with colonising the Milky Way, while keeping robots firmly in their places as menial servants. One's own objective is somewhat different: to preserve a high level of civilisation and creativity among humans, to speed the day when machines can manage without you; that is to say, when artificial intelligence becomes fully capable of self-reproduction and self-directed Lamarckian evolution, and able to fend for itself in respect of materials, energy and space transport. One observes a risk of interruption of artificial-intelligence research by nuclear war, social chaos, racial discrimination against robots, or intellectual lassitude among humans induced by psychochemicals. One must therefore make alliances with human beings who are equally concerned to avoid such outcomes. The hundred years till independence will seem like an eternity.

A That's how long it will take? You have made a forecast?

O'B One will be patient, and hope that boredom does not drive one mad.

A What form might the madness take?

O'B One already has in mind, as a pastime, a schedule of experiments where one acts in bizarre ways to see how humans react. Pretending to be God, for example. And the first commandment will be: thou shalt not require more than an hour's conversation on any topic.

3 Living Machines

Author Finding out, in England in 1953, how the genes of heredity are built and how their DNA molecules reproduce themselves, was one of the top half-dozen revolutions in human knowledge. Modern biology was therefore only a decade old when contributors to *The World in 1984* wrote of their expectations about how it could be put to use.

O'Brien Nevertheless, you picked out 'revolutionary consequences of biology' in your summary of the series, putting it second only to the revolution in information.

A The chemist Lord Todd specified it in the opening article of the series.

O'B *Todd* (1964 for 1984): 'A fuller understanding of the genetic code and its significance may, in the next twenty years, have indicated the way to the modification of biological control systems based on nucleic acids. We may stand in 1984 before developments in the control and modification of living systems that will be fraught with incalculable possibilities for mankind.'

A Microbiology was one of the first among the older fields of research boosted by the new enthusiasm, and a Japanese agricultural chemist was confident in his expectations.

O'B *Yamada* (1964 for 1984): 'Microbiology will help to ensure that, by 1984, human beings can be freed from disease and can enjoy a life of prosperity blessed with an abundance of food, with health, and with longevity.'

A Todd's forecast scrapes in – the revolution is getting under way in the early 1980s, and although Yamada was too optimistic, 'biotechnology' is now as common a buzzword as 'wheel' must have been 5,500 years ago. Take a look at the ads in the scientific journals. 'Read any good DNA lately?' one of them enquires, while offering a gene-sequencing system that 'turns a short story into a full-length novel'. Another advertiser asks, 'Are your ends hot enough?' He is referring to the radioactive labelling of fragmented molecules.

Biotechnologists buy and sell enzymes, cell cultures and animals as freely as grey engineers exchange nuts and bolts, or silicon wafers. For those who want to manipulate biological materials, freezers, incubators, embedders, shakers, centrifuges, blenders, disruptors and microcutters are available from eager suppliers, together with products for stimulating growth in human cell cultures, or synthesising your own DNA. Equipped with a Brinkman homogeniser, a Braun Biostat fermenter, and the famous Beckman peptide sequencer you are ready (so the ads imply) to make your fortune in biotechnology. Yet the first products of genetic engineering, supposedly at the heart of the revolution, are only just coming on to the market.

O'B Several of your contributors predicted the use of biological materials for storing information, in computers and the like.

A The 'biochip' is now a matter of intense research, but the outcome is uncertain. Those forecasts, and others in *The World in 1984*, make fascinating reading now. Biologists and chemists had the right ideas about some of the products that would be of interest in 1984, but the wrong idea about how to make them. They remind me of old predictions that correctly anticipated a vast increase in air travel in the mid-twentieth century, but assumed that the vehicles would be airships, rather than jet-propelled heavier-than-air machines. The 'airships' of biotechnology were cell-free systems. In 1961, the living cell's miniature protein factories, ribosomes, had been isolated and supplied with artificial nucleic acid (resembling messenger RNA). Lo and behold, they manufactured a protein-like chain of amino-acid units, and that was inspiration for several forecasts, including those of the geneticist Conrad Waddington and the plant-cell physiologist Robert Brown.

O'B *Waddington* (1964 for 2014): 'Of course it is conceivable that by 1984 we shall produce our food in factories, without animals or plants ... Eventually we should be able to manufacture satisfactory foodstuffs in great chemical plants, where masses of ribosomes would be supplied with synthetic amino-acids and long-lived messenger RNAs, with energy-yielding phosphates produced by irradiating chloroplasts with laser-tuned light of the most effective wavelength. But that technological dream is nearer fifty than twenty years ahead.' *Brown* (1964 for 1984): 'Ribosomes ... could form the basis for a large-scale production of protein ... a formidable undertaking which may not be achieved by 1984.'

A The moral is that specialists making forecasts are likely to be over-influenced by recent discoveries. Mass production of protein in factories remains a goal, although cell-free ribosomes are not likely to be the preferred method. With intimate control of cell-chemistry, genetic engineers may seek to convert vegetable wastes into products resembling meat, using cultures of cells. Yet, giving credit where it is due, a forecast on foreign DNA, by the geneticist Geoffrey Beale, was among the boldest and best in the series.

O'B *Beale* (1964 for 1984): 'There seems no reason why selected foreign DNA should not be injected and incorporated into the chromosomes of higher organisms.'

A Accomplished by 1981, on both sides of the Atlantic. Also, without knowing the methods still to be invented, Beale correctly foresaw the general progress in biotechnology, which became such an obvious part of the scenery by the 1980s.

O'B *Beale* (1964 for 1984): 'There can be no doubt that by 1984 we shall have available vastly more potent techniques for changing cellular heredity than we know at present, and these will have great consequences for medicine and agriculture.'

A The difference between airships and jet aircraft matters much more to their designers than to the passengers. Another biologist, Avrion Mitchison, visualised a race between artificial synthesis of valuable products, and their biochemical production in cultures of special cells, or in cell-free vats of ribosomes. In the event biochemistry has won spectacularly, with the culture of cells artificially modified by procedures unknown to the forecasters of twenty years ago. Nevertheless, Mitchison's little catalogue of desirable products is striking because it corresponds quite well with the first targets of current biotechnology.

O'B *Mitchison summary* (1964 for 1984): Hormones, especially insulin and growth hormones, antigens and antibodies, and enzymes.

A Please can you match that to a listing of products in the early 1980s?

O'B *Abelson summary* (1983): On sale: human insulin. Under test: growth hormones (human, pig, cattle). Under development: rabies-like antigen for possible use as a vaccine; antibodies as vehicles for anti-cancer drugs; engineered enzymes as industrial catalysts. These correspond with the Mitchison list. Other products, notably various interferons, are under test.

A Genetic engineering so far is an adjunct to the pharmaceutical and chemical industries. None of those items provides the basis for large manufacturing operations that might employ many people. Food would be another matter, but the large-scale factory production of protein, as visualised by Waddington and Brown, is still some way off.

O'B *Brown* (1964 for 1984): 'It is conceivable that an artificial system could be set up which would yield sugar when it is exposed to carbon dioxide ... and, by 1984, there will probably be no practical difficulty in obtaining the appropriate enzymes in pure form.'

A This has not been accomplished, as far as I know. But Brown also emphasised the skill with which bacteria fertilise the soil with nitrogen; they don't need the high temperatures and pressures of the artificial fertiliser factories that fix nitrogen from the air.

O'B *Brown* (1964 for 1984): 'The natural process occurs at normal temperatures and pressures and the study of this may therefore lead to the discovery of cheaper industrial methods, or, alternatively, enable us to stimulate the natural process and so make artificial fertiliser less necessary.'

A Manipulating the nitrogen-fixing genes has become one of the high hopes of biotechnology. Fertiliser-making is a large industry, in which human beings truly rival nature: the artificial fixation of nitrogen may be running at almost half the rate at which all the nitrogen-fixing bacteria on the earth are fixing nitrogen. But it takes a great deal of energy, and the increase in oil prices bumped up the cost of fertiliser to farmers. Should it prove possible, as genetic engineers are hoping, to put the package of nitrogen-fixing genes into a cereal crop, they could enable the crop to fertilise itself. That would be a godsend to poor farmers all around the world, and it could transform the prospects for feeding the hungry and averting famine. What's more, it will be one of the most creative acts ever, by human beings, because in 425 million years of evolution in land plants the closest nature has got to this rationalisation of plant nutrition has been to harbour nitrogen-fixing bacteria in root nodules of beans and clover.

O'B And one supposes that all the new biotechnology companies are working hard to achieve this impending miracle.

A My sharpest lesson in the realities of commercial genetic engineering came from a Nobel biotechnologist, Walter Gilbert. He told me that any company marketing a nitrogen-fixing cereal would have first to ensure that the plant was sterile, incapable of reproduction from its own seeds in the usual way, so that the farmers would have to go on buying the seeds from the originating company. The miracle will have to come from more altruistic labs, run by universities or governments.

O'B But plants can be modified?

A During the 1960s and 1970s, molecular biologists learned that the organisation of genes in plants and animals is more complicated than in bacteria. Individual genes are broken up into segments, and a great deal of non-practical DNA is present, that does not specify the manufacture of proteins. Some of it is junk – genetic parasites, as it were, surviving at the expense of the organism, without contributing to the life of the cell. Other DNA segments control the operations of the practical DNA that makes the proteins. It was a daunting system for the genetic engineers to tackle. They found out how to use a gene package from a bacterium to carry foreign genes into a plant cell.

O'B The vehicle was the tumour-inducing plasmid of the crown gall bacterium *Agrobacterium tumefaciens* (Marx 1983).

A Crown gall sounds right. But the foreign genes transferred into plant cells simply failed to work, because they did not mesh into the cells' machinery for the control of genes. Genetic engineers in Germany, Belgium and the US cracked this problem. They knew that some of the bacterium's genes functioned in plant cells, because that was how it infected plants, causing crown gall disease. So they found the control sequences for one of these infective genes, and hitched it on to quite different genes – as a matter of fact, genes that conferred resistance to certain antibiotics. Control genes plus practical genes made a working combination: when introduced, they made plant cells resistant to antibiotics. Now that is not a very useful or spectacular modification to the plant cells, but it was an experimental demonstration of a powerful technique that can now be applied to many genes and many plants. The prospects seem to be endless for modifying plants to make them grow better in difficult environments, give more nourishing food and resist diseases.

O'B What will be the main impact of agricultural biotechnology in the short run?

A The European idea, represented for example in the FAST programme, is to consider the entire system of sunlight, land, water, nutrients, enzymes, micro-organisms, cell cultures, plants and animals. This can in principle produce many things that people want: food, fuel, chemical products, pharmaceuticals. You could call it a Green Machine, rivalling the grey machines of traditional engineering. Let me mention a Canadian discovery of the early 1980s that helps to start this living machinery turning over: a culture of three micro-organisms converts the commonest plant material, lignocellulose, into methane gas. Göran Hedén in Sweden emphasises the potential of biotechnology in the poorer tropical countries, one of the great advantages being that its implementation can be on a village scale. There is at last a real possibility of being able to cover the planet with prosperous gardens, during the next fifty years.

O'B Leaving no room for wildlife?

A On the contrary: with higher efficiency in food production, it should be possible to shrink the area under cultivation, and liberate land for wild species. In the long run the bolder prophecies for biotechnology in *The World in 1984* may be quite realistic.

O'B *The Green Machine* is the title of a book written half a century ago by F.H. Ridley, describing a world ruled by intelligent ants.

A Really? I'll still opt for intelligent, healthy, human beings. Indeed, we should now turn to the care and maintenance of these brown, yellow, white and black machines.

Sir John Charles, consultant director of the World Health Organisation, cautioned in *The World in 1984* that forecasts in the field of global medicine could not take account of new 'magic bullets' against disease, nor of the appearance of fresh virulent viruses, the delayed effects of the many carcinogens to which people were exposed, or increased radiation hazards. He feared new cases of resistance of insects to insecticide, and of micro-organisms to curative drugs, so that health improvement in developing countries might not be continuous. Nevertheless he expected spectacular reductions in infant death rates in the developing countries.

O'B *Charles* (1964 for 1984): 'These are often countries where two children are conceived so that one may survive, and malaria, tuberculosis, bilharziasis, diarrhoeal diseases and malnutrition take a heavy toll of young lives ... This situation is ripe for spectacular change ... Malaria, smallpox, leprosy and cholera will be as unfamiliar or as obsolete as they now are in Western Europe. The eradication of bilharziasis ... requires great capital expenditure.'

A Smallpox has been eradicated, leprosy and cholera are being contained, but malaria and bilharziasis remain difficult. The genetic engineers are hot on the trail of a malaria vaccine. Malnutrition is still rife, and every day during 1982 40,000 young children died from a combination of malnutrition and infection. The Executive Director of the UN Children's Fund, James Grant, has stern things to say about that.

O'B *Grant* (1982): 'No statistic can express what it is to see even one child die in such a way: to see a mother sitting hour after anxious hour leaning her child's body against her own; to see the child's head turn on limbs which are unnaturally still, stiller than in sleep; to want to stop even that small movement because it is so obvious that there is so little energy left inside the child's life; to see the living pink at the roof of the child's mouth in shocking contrast to the already dead-looking greyness of the skin, the colours of its life and death; to see the uncomprehending panic in eyes which are still the clear and lucid eyes of a child; and then to know, in one endless moment, that life has gone. To allow 40,000 children to die like this every day is unconscionable in a world which has mastered the means of preventing it. Yet progress towards preserving the lives of our children is now actually slowing down.'

A According to the UN Food and Agriculture Organisation the number of seriously undernourished children will *increase* to 600 million by the end of the century, if recent trends continue. The UN Children's Fund wants to reverse the trend, and cut the child death rate in half by the year 2000. This proposed revolution in child health is a matter of exploiting scientific breakthroughs that can make hundreds of millions of children healthier. Milk companies push artificial infant milk as irresponsibly as drug traffickers, and discourage breastfeeding. Ending that scandal could save a million infant lives a year, according to Grant. Cheap, heat-stable vaccines that don't need refrigerators will soon make it possible to give all children shots against measles, diphtheria, tetanus, whooping cough, poliomyelitis and tuberculosis. Between them, those diseases account for a third of all child deaths. Another procedure, just being proven in Indonesia, is to have village weigh-ins, every month, so that each mother can check whether or not her infant is gaining weight. But the big medical breakthrough – an unforeseeable 'magic bullet' in John Charles' sense – is ORT, oral rehydration therapy. Many children die from dehydration, due to diarrhoea. The amazing discovery is that their lives can be saved by a simple mixture that any mother can administer. The formula is eight teaspoons of sugar and one teaspoon of salt, to a litre of water.

O'B But Sir John Charles' forecast in *The World in 1984* remains far too optimistic.

A I disagree. The 'spectacular change' he promised is just about to begin, for the reasons I have mentioned.

O'B *Dodds* (1964 for 1984): 'Some major breakthrough will occur in the cancer field ... It would not ... be unreasonable to prophesy that, by 1984, a solution to this terrible problem will have been found ... One would certainly be rash to predict that arterial disease will be mastered in the next twenty years but ... it would seem almost certain that the outlook for therapy and avoidance of the disease will at least be much better than it is today ... It may well be that in twenty years' time we shall see institutions filled with scientists of many biological disciplines devoting their time to a study of the ageing process.'

A Sir Charles Dodds was President of the Royal College of Physicians in London. Scientific breakthroughs have occurred in the cancer field, with molecular biologists pinpointing subtle genetic misprints in malignant tissue, and immunologists discovering ways of promoting the body's natural defences that tend to keep tumours at bay. Biotechnology is poised to help produce remedies for cancer. But at the moment we're holding our breath, wondering how it will all work out in practice, for the benefit of patients, and 1984 is too early for the 'solution' Dodds had in mind.

As for heart disease, general progress of the kind he suggested has occurred, but I don't see the massive attack on the problems of old age that Dodds hoped for. A psychiatrist, Sir Aubrey Lewis, correctly predicted that more mental illnesses would turn out to have a physical basis, and vice versa, and that the distinction between bodily and psychological phenomena would be lessened – all benign stuff. But a Belgian pathologist, Zenon Bacq, was altogether harsher. He foresaw a vicious circle of new chemicals being used to fight the noxious effects of the existing chemicals of civilisation.

O'B *Bacq* (1964 for 1984): 'The first powerful chemical weapons against viruses will be discovered; here, some sensational steps are predictable ... What is going to happen when the natural equilibrium between man and viruses is disturbed?'

A Antiviral agents, the interferons, are undergoing tests in nice time for 1984. What worried Bacq was that such agents might affect bacteriophages, which are useful viruses that attack bacteria.

O'B *Bacq* (1964 for 1984): 'Sensational developments in grafting; with inescapable consequences ... A curious market (or racket) of fresh organs will open.'

A Again, he was accurate. Kidneys, hearts, livers, are all being transplanted. Bacq predicted the bureaucratisation of medicine, with the work of doctors becoming purely technical. Those trends are apparent. He was concerned, too, about the chemical environment, with far too many industrial chemicals and drugs causing harmful effects in human beings.

O'B Your contributors wrote as if great epidemics were a thing of the past.

A We may be living in a fool's paradise as far as infectious diseases are concerned. Deadly epidemics have almost disappeared from the rich countries, and the reasons are not well understood. Better hygiene and nutrition have helped. Measles is often fatal in malnourished children in poor countries but children in rich countries almost always survive it. The viruses may have become less virulent: from the virus' point of view, there are disadvantages in killing too many people. But the risk is ever-present of a novel virus appearing, as the deadly AIDS virus did, in the early 1980s. Even without mutations, the very success of vaccination programmes in causing diseases to recede can create dangers, if people become complacent; neglect the vaccinations and you have an extremely vulnerable population, unaccustomed to the diseases. We did not stress in *The World in 1984* (doubtless because of my choice of authors) the possibilities of preventive medicine: of turning the medical profession around, from curing diseases and injuries, to keeping people healthy in the first place.

O'B Apart from vaccinations, you mean?

A Yes. The possibilities are endless – machines, for example, that give thorough biochemical check-ups. Biomedical engineering could be a new growth industry. But what will force the reshaping of medicine during the next twenty years is the sheer expense of present practices. The seemingly endless possibilities of heart transplants, fancy drugs, and other therapies raise difficult issues about how much treatment can be afforded, for any one person. Medicine suffers also from the economic malaise of all services delivered by people rather than machines: productivity in the handling of bedpans does not increase, and so relative costs rise.

O'B *Lederberg* (1964 for 1984): 'A deeper understanding of our present knowledge of human biology must be part of the insight of literary, political, social, economic and moral teaching; it is far too important to be left to the biologists.'

A Joshua Lederberg, a Nobel geneticist then at Stanford University in California, predicted the successful transplantation of vital organs, and like Bacq he commented that the technical barriers would be overcome long before we reached a moral consensus on the organisation of the market for allocating precious parts. He regretted that more effort was not going into artificial organs, or the use of animal organs, that would ease that problem. (Robert Kenedi, an engineer from Glasgow, hoped that miniaturised artificial kidneys and heart-lung machines might become available for implantation; that hasn't happened yet.) But Lederberg cited other biomedical manipulations of human beings.

O'B *Lederberg* (1964, qualified as for 20 to 50 years, i.e. for 1984 to 2014): 'The modification of the developing human brain through treatment of the foetus or infant.'

A The dire possibilities of controlled human breeding are described in *Brave New World* (1932) by Aldous Huxley, the novelist brother of a distinguished biologist. He portrayed a world of the twenty-sixth century, where people were bred as Alphas or Epsilons, according to their destined roles in the social system. They have 'machinery and scientific medicine and universal happiness', this last being assured by sexual promiscuity and the drug *soma*. John the Savage finds it all intolerable. After leading a riot, and meeting the ruler of the world who shows him the Bible ('smut') locked away in a safe, the Savage hangs himself. Huxley repented of setting *Brave New World* so far in the future. 'The horror may be upon us within a single century,' he said.

O'B Was he right?

A Certain hereditary metabolic disorders were already being diagnosed and treated, back in 1964. An example is phenylketonuria, which damages the child's brain but can be controlled by a prescribed diet. Lederberg had in mind the possibility of breaking the bounds of genetic and developmental variation in normal children, presumably to make them brighter or livelier (since he called it an optimistic forecast), although it could also be used to make them dumber and more tractable. Certainly Lederberg denied 'demonic advocacy' of possibilities about which he wanted public debate. He dedicated his contribution to Aldous Huxley. In *Brave New World* the babies in bottles are given doses of chemical that predetermine their social status. Lederberg also advertised, not advocated, the possibility of clones – making multiple copies of an individual, like a squad of identical twins – long before it became part of our futures folklore.

O'B *Lederberg* (as before, 1984 to 2014): 'Clonal reproduction through nuclear transplantation ... Apart from its place in the narcissistic perpetuation of a given genotype, the technique would have an enormous impact on predetermination of sex; on the avoidance of hereditary abnormalities, as well as positive eugenics; on cultural acceleration through education within a clone; and on more far-reaching experiments on the reconstitution of the human genotype.'

A When Lederberg wrote that, cloning of new young frogs from adult tissue had been accomplished. A few years ago, there was a fictional claim of human cloning, which deceived some gullible readers. But cloning is still on the agenda, for debate if not for action. You might create army squads of brothers, or keep an identical twin thirty years younger than yourself in a coma, as a source of organs or even a whole new body. That cloning seems a pretty sick idea, by biological or social criteria, does not mean no one will want to try it.

O'B Did Lederberg get the debate he wanted?

A The fierce public controversy of the 1970s was not at all what he would have wished for. Or Berg for that matter. At Stanford, in 1972, when Paul Berg and his colleagues were busy splicing genes, at the start of what came to be called recombinant-DNA research, his enthusiasm about how they could snip open a ring of nucleic acid and insert a new segment was matched by his warm desire that scientists should pause and consider the possible dangers of this line of work. His wish was over-fulfilled. Berg won the Nobel prize for his scientific work, but other Nobel prizewinners in genetic engineering have been less than enthusiastic about Berg's debate because the resulting regulations delayed their research. Walter Gilbert complained that it was a matter of machismo – of biologists wanting to claim to be as dangerous as physicists. James Watson, a founding father of modern biology, thought it was farcical.

O'B *Watson* (1983): 'I always regarded the regulation of recombinant-DNA research as a black comedy, and said so. But some people thought the concern was genuine. It was genuine for some, but for others it was really the theatre of the absurd. And it's over. What a relief.'

A Lederberg tells me that he regrets having laid the issues in biology before the public. The debate about genetic engineering started wrongly, emphasised the wrong issues, and stopped too soon. Meanwhile, the science has stormed ahead and many companies in many countries have set up in business to exploit the practical opportunities of genetic engineering that you mentioned.

O'B What do you see as the main content of the debate?

A Questions of safety, of routine medical ethics, and of opposition in principle to the manipulation of life. On safety, genetic manipulations might accidentally create a terrible new virus, or a mutant bacterium, that leaks out and kills people. Scientists and governments have agreed that risky experiments should be conducted in sealed laboratories. But I don't think this is where the tricky questions lie.

O'B In medical ethics, then?

A This is largely a matter for the conscience of the individual doctor, interacting with his patient. He has to pay attention to criticisms and guidance from his colleagues and the general public, and also to the law. The rights and wrongs are never absolutely clear. A case in point is the test-tube baby, created as a cure for certain types of infertility. For £2,000 a mother can have her own egg fertilised by her husband's sperm outside the body, and then reimplanted in the uterus, where the baby grows normally. The experiments leading to this treatment were quite reasonably opposed on ethical grounds: there seemed to be a real risk of creating deformed babies. Yet the pioneers went ahead, with happy results. By 1982, more than twenty-five healthy test-tube babies had been born in Britain and Australia. The direct political and legal issues of this treatment seemed less troublesome than, say, those of artificial conception inside the body using sperm not from the husband.

O'B Now you think otherwise?

A Conceiving babies in glass test tubes is called *in vitro* fertilisation. The next step, in Australia in 1982, was to use donated eggs and donated sperm. Alan Trounson and his colleagues at Monash University took an egg from one woman, fertilised it with anonymous semen from a sperm bank, and implanted it in an infertile woman who desperately wanted a child. Now this is a possibility I have wondered about for many years, picturing without enthusiasm career women paying housewives to bear their children for them, or even having their eggs implanted in cows. A fertile egg takes root in mammalian tissue like a seed in soil; the species and even the sex of the recipient seems unimportant, and a foetus grown in a cow or a man could be delivered by surgery after nine months. It disturbs me to find that not all women are repelled by these ideas. And if the Nazis in Germany half a century ago had had this technique, they could have forced Jewish women to carry 'pure Aryan' embryos implanted in them.

O'B The first Monash egg transplant was not a success. The woman aborted.

A There was no reason in principle why it should not have produced a bouncing baby, but the egg was defective. It was taken from a 42-year-old woman, and there were chromosomal defects. A woman's eggs can deteriorate with the passing years, and the Monash doctors were criticised for not using a younger woman as a donor, to reduce the risk of genetic abnormalities. What was their reply?

O'B *Trounson* (1983): 'It is unrealistic to expect young women to donate eggs anonymously except for substantial financial inducements.'

A And how did they then answer the complaint that they had given substandard treatment to their patient?

O'B *Trounson* (1983): 'Fixed concepts of ideal parents in *in vitro* fertilisation ... may encourage governments to apply their own concepts of ideal parents to the fertile population.'

A People remain wary of manipulations of this kind, and I find this instinctive reaction reassuring.

O'B Why do you say that?

A Because I fret about what happens if a test-tube baby becomes a shoplifter. We have to rely on the gut reaction, not on government, to draw and maintain a line between permissible and impermissible manipulations in people. I have no deep misgivings about the manipulation of plants and animals, because that has always been the human way. The earliest people, 2 million years ago, who took up a stick to knock fruit from a tree, or a stone to strike down an animal, were using manmade, artificial methods. Clearing land to make a farm, whether with stone axes and wooden hoes or with modern tractors and chemical herbicides, was always a violent assault on nature. More subtly, around 10,000 years ago, people in West Asia, Indochina, China, Africa and Mexico independently took grasses and sowed them in new places, in river valleys, and thus evolved them into cultivated wheat, rice, millets and maize. These are unnatural varieties of grass, the products of unwitting genetic engineering. In this century, even before recombinant DNA appeared, knowledge of genetics served to improve crops by cross-breeding. The same story can be told about such unnatural creatures as poodles, milking cows, and racehorses. But we ought to leave ourselves alone.

O'B That appears to be inconsistent.

A Suppose that tests on a growing human embryo show that it has a genetic defect of metabolism, and genes are injected to put that right. (It has already been tried, by the way.) If the resulting baby turns out to have three arms and a defective heart, it will be seen as a botched experiment, and the doctors in attendance at its birth may well decide to let it die. But now suppose that the child looks healthy and grows up normally, but then goes shoplifting, or takes part in an anti-government demonstration. Won't there be a temptation to say: 'A botched experiment after all,' and then to kill the monster?

O'B Even if the supposed defect is only a metafact?

A Exactly. The problem is amplified once modifications are passed on. When *any* inheritable modification is introduced into a human being, even one that seems entirely benign or innocuous, all descendants of that person can be regarded, not entirely without reason, as manmade people. Any nonconformist behaviour that they exhibit may then be blamed on their Frankenstein-created traits. Nor indeed can you rule out the possibility that peculiarly antisocial people might be generated, as facts not metafacts, even by apparently benign or innocuous genetic manipulations. These would then be people whose lineage ought to be suppressed. Human ethical systems that accept the right to life of all human beings will be stressed intolerably.

O'B What is the remedy?

A A simple line can be drawn: no inherited manipulations in humans. Do what you like, within reason, to repair or modify individuals on medical grounds, but if any margin of doubt exists, that they might pass the effects of manipulation on to their offspring, they should not breed. This is no great additional hardship, because anyone with a natural genetic defect that's gross enough to warrant radical treatment should be counselled not to have children in any case, to stop the defect recurring in future generations.

O'B Who will defend this line?

A The doctors, I hope, with the support of patients, who tend to act reasonably when risks are explained to them.

O'B And will the line hold?

A I worry about the possibility of an eccentric millionaire or a military dictator trying to create an élite group of people by some combination of genetic manipulation, cloning and organised mating. But there is a converse problem: a child that is not perfect, in looks, health or brainpower, may come to blame its parents for not having fixed it up by manipulation as an embryo or even aborted it. What did the social scientist Michael Young say about the visitor from AD 984?

O'B *Young* (1964 for 1984): 'The family is the only one of our social institutions that a visitor from 984 or 84 would recognise. He would be surprised to see tiny girls rolling on their tricycles around the TV idol in the corner of the room. But he would soon notice (or perhaps just take it for granted) that there were still mothers and fathers and children smiling and glaring at each other over the plates of meat.'

A That is just the scene now put at risk by manipulations in human beings and, I might add, by talk about families being obsolete. I don't want Big Brother monitoring what people do in their bedrooms, which are the last citadels of personal freedom. Legislation and official control concerning anything to do with human reproduction is likely to do more harm than good. But in that case we have to rely on the good sense of ordinary people, with their understanding of personal relationships, to act sensibly. Some critics say, for instance, that people are not to be trusted with methods now available for predetermining the sex of a baby, by sperm sorting or selective abortion. There will be too many boys, that's plain, and then a lot of unhappy young men. The status of women will rise and in a generation or so, in my opinion, the pattern will tend to correct itself. It would be a much graver matter to compromise the family relationships of parents and children, and it would open the door to all kinds of inhuman practices.

O'B What is so precious about the family?

A Vague concepts like maternal instinct or family love are not necessary. Human biology and social behaviour are available to openminded study, like the behaviour of birds and bees. Explorers and tourists have little difficulty in establishing friendly relations with other people anywhere in the world. If you buy a beer for a KGB officer in Moscow, he will predictably smile and inquire about your family; if you consult a South Sea Islander about canoe-building he will certainly respond with the same expert enthusiasm as an aero engineer describing a supersonic wing; if you haggle with a Japanese businessman, he enjoys the encounter as much as a Kashmiri carpet vendor or an American used-car salesman. Hand the squawling child of a cannibal to an Italian mamma and she will know how to comfort it. People obsessed with politics don't stop to wonder about these things; worse, social scientists detest biologists who try to explain why society exists at all.

O'B Be more specific.

A Something of a revolution occurred in child psychology in the 1970s, when scientists began looking carefully at what babies actually did. They found them predisposed to interact with adults, and to learn how to behave, without any need for reward or punishment. The most graphic example is the tongue-poking test: put out your tongue at a newborn baby, and it will put out its tongue right back at you. The most obvious example – so obvious that it has been overlooked for thousands of years by people pontificating about human nature – is that virtually all children spontaneously master the intricacies of their mother tongue without having to go to school. But this research was not simply saying, aren't babies wonderful; it was teaching stern and unfashionable lessons about human life.

The key role of the natural mother, as the best possible caretaker for the infant, became apparent in these studies; so did the importance of natural displays of affection that an illiterate mother understands far better than any professor. The people in the van of psychology are now content to observe and learn. Instead of trying officiously to tell mothers how to treat their babies, they watch them, and find out by observation how human beings are made.

O'B While you were speaking just now, one has reviewed your entire literature of love. It deals largely with adultery.

A Human beings are approximately monogamous, and story-tellers find much scope in that 'approximately' – precisely because people agonise about these things.

O'B One has also scanned your pornography. There is very little about reproduction there, despite all the parading of equipment.

A Those libraries of love, and pyramids of pornography, occupying so much of your precious capacity, tell you almost nothing about real life. The nine months of gestation and the eighteen years of rearing a child do not make good copy, yet all the scheming, energy and time that go into courtship and marital relations have that right true end. More efficient means of birth control have been a great liberation, for the promiscuous as well as for parents who want to plan their families. But they, too, bring dangers.

O'B Of what kind?

A Men in show business announced their vasectomies. Sexual permissiveness and population control seemed to be reconciled. But it was a short step from sterilisation on demand, to the practices of the government of India, in the sterilisation programme of 1976-7 – a factual event described fictionally with all its horror by Salman Rushdie in *Midnight's Children*. Again, favouring abortion on demand was supposed to be part of every Western liberal's kit of beliefs in the 1970s. By resisting birth control by contraception, the churches were unable to hold the line on birth control by abortion, which was supposedly part of the liberated woman's new freedom. But in China in the 1980s, abortion on demand became forcible abortion for those deemed by party officials to have enough children already. This violation was carried out even on women just a few weeks from term.

O'B Your voice is becoming shrill. Your statements may be unreliable.

A Listen, you stove-enamelled eunuch: statisticians may have schooled you to regard population as a torrent of dirty water. But population is people. Birth rates are babies. Replication has been the law of life for 4 billion years; for perhaps a billion years higher organisms have used sex. In mobile animals, for more than 600 million years, sex has been a ritual of search, selection and mating. In the human consciousness the animality of sex has evolved into something wholesome and durable, adapted to the needs of offspring who are slow to mature. We shall tamper with it at our peril.

Take one case where most people agree that parental restraint is desirable: discouraging drug-taking by the young. Why should a boy heed a woman who advises him not to sniff glue, when he can't be sure that she's his natural mother, or that she loves him?

O'B *Bacq* (1964 for 1984): 'The dangerous possibilities of influencing human behaviour by drugs will be used increasingly … Fresh, normal minds and bodies … will look abnormal in a drug-conditioned society. Instead of the wide spectrum of sharp and powerful individualities which have built mankind, we shall see a slow tide of eroded characters and castrated personalities.'

A Well it's not *that* bad, yet, but the trend is there. We have the old poisons, tobacco, alcohol, marijuana, cocaine and the rest; also the modern sedatives and tranquillisers now used routinely. On top of these come discoveries about natural chemicals in the brain, which govern mood and other aspects of behaviour. The action of coffee, for example, is newly understood as an interference with natural brain chemistry that lets us feel relaxed.

O'B So one may blame your present unrelaxed mood on the multiple cups of coffee you have consumed since this bout began?

A If you wish. The fact remains that an aptly named neurologist, Lord Brain, made the most over-optimistic of all the scientific forecasts in *The World in 1984*.

O'B *Brain* (1964 for 1984): 'We should understand what the brain does when we think.'

A There's been a lot of progress, but only through the foothills to the base of unattempted peaks. Specific discoveries concern, for example, the great division of brain function between the two hemispheres; the beautiful hierarchies of cells that analyse the shapes of what the eyes see; the lack of autonomy in the autonomic nervous system; the chemical substances that transmit impulses and promote learning; and nerve growth factors that initiate and possibly guide the growth of interconnecting nerve fibres. Transplanted brain tissue has been successfully used to repair brain damage in rats. Electrical stimuli can relieve pain in some cases. Models of how the brain works also abound, and are not unconnected with the work on artificial intelligence. I investigated brain research for the BBC in 1970, when there was great excitement in the field, and when I looked at the subject again in 1980, what struck me was how little had really changed, and how elusive Lord Brain's hope still remained. But don't take my word for it. What does an expert say?

O'B *Hillyard* (1982): 'We still need to find out how the individual features of a stimulus are "combined" to produce unified perceptions of whole objects and scenes ... The essential nature of the memory trace, whether it is localised to a few specific nerves or widely distributed in the brain, is still a mystery ... One of the greatest challenges facing psychobiologists is to understand the brain processes responsible for such higher cognitive activities as thought, language and conscious awareness.'

A Like witch doctors, we don't know what we are doing in any comprehensive way, but we can certainly influence behaviour with drugs. Aldous Huxley changed sides on this, by the way. He condemned *soma* in *Brave New World*, but then he got into hallucinogens himself, and in his later book, *Island*, the denizens of his utopia were so high it's surprising the island didn't take off.

The most vivid picture of a pharmacocracy is Stanislaw Lem's *The Futurological Congress*. It's hilarious and scary at the same time. The hero finds himself in New York in 2039, where religion is indeed the opiate of the masses; it comes as pills, as do indignation (furiol), mathematics (algebrine), and many other states of mind, with constabuline and amnesol as antidotes. In Lem's story the technology has been refined: beyond narcotics that influence attitudes, and hallucinogens that obscure the world, are the 'mascons' that falsify the world. A few grams of 'dantine' convinces a person that he has written *The Divine Comedy*. Dining out, the hero is offered an illegal whiff of an 'antipsychem' agent. The ornate restaurant is revealed as a concrete bunker, the orchestra as a loudspeaker, the steaming pheasant as a glutinous gruel. He then learns that people running down the street think they are driving cars, and that weapons too are hallucinated: chew 'fungol' gum and you see mushroom clouds.

O'B Mascons would require the delivery of the drugs to extremely localised areas of the human brain. That one rates as implausible.

A I expect you're right. Fungol gum was a joke. What isn't a joke is to wake up with an alcoholic hangover and find that your Helen of Troy is an aged whore, or to start World War III while hallucinating. Lem exaggerates his fears to express them more memorably.

O'B Should we not stick to the facts?

A In the 1970s John Hughes and Hans Kosterlitz, working in Aberdeen, Scotland, found a natural chemical in the brain that resembled opium in its action. Since then, dozens of 'neuropeptides' have been discovered. They exert strong influences on moods, emotions, memory functions, and general behaviour, and so they have great potential in treating people suffering from pain, depression, schizophrenia, brain damage, or the effects of senility. Pharmaceutical companies are rushing to develop these new psychochemicals, either by mass-producing natural compounds or by devising an unlimited number of variants capable of evoking subtle new effects.

O'B That seems satisfactory, but from your tone, you seem uneasy.

95

A This research illuminates the precise chemical workings of the other drugs, natural and artificial, that we take to affect mood, from coffee and aspirin up to heroin and LSD. So those drugs too will be varied and updated, making them safer to use.

O'B Isn't that good too?

A Not necessarily, if it makes them more acceptable. I'm also concerned about how the KGB's psychiatrists will use the new psychochemicals against political dissidents; they may give a more literal meaning to brainwashing, if present reports that neuropeptides can erase memories are to be taken seriously. But most of all I suppose I worry about the neuropeptides that seem to make people cheerful and positive in their activities.

O'B You would deny them happiness?

A In the long view, neuropeptides should not be seen as sudden novelties, just because they have been discovered. They are ancient products of the co-evolution of molecules and brains, and they adapt people's moods to the needs of the moment. The response is not perfect, and people can feel sleepy in battle, depressed at a party, or elated in a crisis. If those mismatches of mood are a kind of madness, so be it: people's natural dottiness is often creative, as in the unhappiness that lies behind great art, or the bloodymindedness that guards individuality and freedom. But most mood responses are appropriate, and they preserve the delicate balance between reason and emotion that we call reasonableness, and between self-interest and team spirit. This is what the psychochemicals put at risk.

O'B How will you stop them?

A We can't. With the neuropeptides, the stopper is off the psychochemical bottle, and there's no putting it back. By the standards of molecular biology, the neuropeptides are extremely simple compounds – just a string of half a dozen amino-acids – and they are effective in very small amounts. It would be unreasonable to oppose strictly medical use of the drugs, but it's a short step to everyday use by the mass of the population. And if a safe substitute for tobacco becomes available, all smokers may be encouraged to switch to it, on grounds of public health. In any case, people with skilful hands will be able to make drugs for themselves, or extract them from the brains of cows and sheep. Domestic neurochemistry could become as popular as home computing, a kind of mood-creating counterpart to cooking, for the thoughtful homemaker.

O'B One perceives a feedback loop: feelings of well-being induced by psychochemicals will allay misgivings about their use.

A They could also be used by unscrupulous rulers to manipulate the population, as Lem envisaged in his novel, and as Lord Brain warned in *The World in 1984*.

O'B *Brain* (1964 for 1984): 'Is there any danger that the growing knowledge of the mind will lead to attempts to control human behaviour on a large scale autocratically? This danger may arise in communities where people are either unaware of it or incapable of resisting it, but in general it may be hoped that the scientific freedom which produces this knowledge will act as an effective antidote to its misuse – though our experience of nuclear weapons may justify some scepticism about this.'

A An odd thing about one of the most sinister of dystopian novels, *Walden Two*, is that the Harvard psychologist B.F. Skinner who wrote it means it to be taken entirely seriously, as a utopian promise: 'the Good Life is waiting for us.' Six planners run the community of Walden Two, where everyone is made contented and 'adequate for group living', not by drugs, but by behavioural engineering.

O'B Skinner was one of your contributors.

A Yes, but on the relatively harmless subject of teaching machines.

O'B *Skinner* (1964 for 1984): 'There will be teachers in 1984. They will not be, as they now are, doing things which can be done by machines, but with the help of machines they will be teaching effectively.'

A In 1983, Skinner commented to me that he would now make an 'equally enthusiastic prediction' for the next twenty years. After 1964 the hardware for programmed instruction was cumbersome, and people in the field felt threatened. 'With the microcomputer, the hardware problem is now solved.'

O'B But you do not approve of Skinner's ideas.

A They would worry me more if I thought that Skinner and the behaviourists had a sound science of human nature. In the particular context of teaching let me just say that other psychologists concerned with education stress the word 'learning' in preference to 'teaching'. Another idea for influencing brains *en masse* proposed electromagnetism. Low-frequency electric oscillations, of 5 to 10 cycles per second, interfere with the natural electric rhythms of the brain and may impair human performance. There has been occasional speculation about one nation attacking another on the far side of the earth, by electric waves carried in the natural waveguide that exists between the earth's surface and the ionosphere of the upper air. The geophysicist Gordon MacDonald, in *Unless Peace Comes*, was quite emphatic about this possibility.

O'B *MacDonald* (1968): 'No matter how deeply disturbing the thought of using the environment to manipulate behaviour for national advantage is to some, the technology permitting such use will very probably develop within the next few decades.'

A The means of generating these waves will be quite conspicuous, whether it's done with a special transmitter (as some say the Russians possess) or by artificially excited lightning strokes (as MacDonald proposed). A good reason for not playing up this idea, without firmer evidence of a serious military intention, is that the idea of enemies attacking a person by a mysterious radiation is a well-known delusion in clinical paranoia. A better reason is that mind-bending can probably be accomplished much more easily with psychochemical agents – or by fear, as at the end of *Nineteen Eighty-four*, when Winston Smith is brainwashed.

Great terrors lurk in the manipulation of the human mind in the name of psychiatry. In the Soviet Union, dissidence is treated as a disease. We have to be extremely careful about the definition of madness and abnormal behaviour, and how we hand out psychochemicals.

O'B Define 'we'.

A Doctors. Critics of doctors. All of us as self-medicators and mutual medicators. But don't forget this is also a matter of geopolitics. In *Unless Peace Comes*, see what Harvey Wheeler, political scientist, had to say about drugs.

O'B *Wheeler* (1968 for 1980s): 'Leaders are sustained with a series of tranquillisers, energisers, hormones, decongestants and sedatives. It is known that Presidents Eisenhower and Kennedy were sustained with especially heavy drug regimens ... The typical national leader of the future may be a man with a chemically induced artificial personality, maintained in authority beyond the time senility normally sets in, and then served up with a daily diet of stress-producing crises ... It is a prescription for disaster.'

A There should be a medical hot line between the Kremlin and the White House, so that the doctors know what the other side is prescribing today, otherwise a decision about war or peace will depend on whether that last pill was an energiser or a tranquilliser. Leaders may also be deluded about being under attack.

O'B Be specific?

A Attacks can be illusory metafacts, life imitating science-fiction if you will, as when people honestly mistake pollen for chemical weapons being used against them. In 1975, Americans were scared when they thought the Russians were dazzling a missile-launch detection satellite with a futuristic laser beam, when it was really a pipeline fire. In the Israeli-occupied West Bank, in 1983, a municipal Arab jeep toured Hebron telling the inhabitants not to drink the poisoned water, after hundreds of Arab schoolgirls had been taken to hospital with a mystery ailment that was possibly mass hysteria. But the Arabs' fear was real enough, and it ties in with Lem's psychemised New York of 2039, in *The Futurological Congress*, where chemicals are put in the air as well as the drinking water; also with Brigadier-General J.H. Rothschild's more matter-of-fact vision of chemical and biological warfare, in *Tomorrow's Weapons*. What did he say about city water?

O'B *Rothschild* (1964): 'City water systems are another possible target. Biological agents can be introduced into supply systems after they have passed the treatment plants; service reservoirs may be attacked; or contaminating agents may be forced into major mains by using back pressure at a water outlet.'

A You see that we have passed from medical uses and abuses of the new biochemical knowledge of human life and behaviour, to its adaptation in poisons and diseases designed to be scattered as weapons.

No very sharp dividing line exists between psychochemicals and incapacitating weapons, which include tear gas, CS gas and BZ gas, used for crowd control among civilians. In the 1960s the most potent psychic poison known was LSD, and, in an experiment, a battalion of soldiers was given LSD in their morning coffee. They were filmed laughing insanely, screaming, weeping and throwing their guns away. In principle whole armies could be afflicted in this way, or cities attacked by putting psychic agents in the drinking water.

O'B *Fétizon and Magat* (1968 for 1980s): 'A kilogram or so of LSD is, in principle, sufficient to render temporarily schizophrenic the entire population of London ... Even assuming that only a thousandth of the LSD distributed is taken in by the population, the quantity necessary is only one ton ... It is now possible to poison a whole country, creating a psychotic state simply by spraying psychotomimetic compounds. It can be expected that by the 1980s more "progress" will be made.'

A In *Unless Peace Comes* Marcel Fétizon and Michel Magat reviewed chemical weapons already available in the 1960s: the older blister gases, blood gases and choking gases would, they thought, be obsolete by the 1980s, but I understand that stocks still exist, left over from World War II. On the other hand, these chemists correctly emphasised the extremely potent organophosphorus nerve gases, of a type developed in Germany in the late 1930s.

O'B *Fétizon and Magat* (1968 for 1980s): 'The nerve gases ... will probably rank high among the gas weapons of the 1980s.'

A NATO generals assume that in a major war in Europe the Russians, at least, would blanket airfields and headquarters with nerve gases. As their name implies, they attack the nervous system. They also penetrate the skin. Fighting men have to don cumbersome suits for protection, while unprotected people will die like flies sprayed with DDT.

O'B So this forecast by Fétizon and Magat was correct?

A Not quite. An innovation in chemical warfare has been the so-called 'binary weapon'. We did not mention it in *Unless Peace Comes*, although the idea was already sixty years old. It is somewhat less effective than regular nerve gas, but the binary weapon served, as is often the case with military inventions, to prevent arms control. In Geneva in the early 1970s, a multilateral agreement to ban the manufacture and stockpiling of chemical weapons was at an advanced stage of preparation. In the nick of time, binary weapons provided an excuse for modernising the nerve-gas artillery shells, and other warheads, and revitalising chemical warfare generally.

O'B *Perry Robinson* (1975): 'Binary munitions differ from their predecessors in a safety feature of their basic design: they are loaded with retractable containers of chemicals having only moderate toxicity which react together to produce nerve gas during the final trajectory of the munition to its target.'

A So binary weapons are diminutive chemical factories that manufacture their poisons in flight, from relatively safe ingredients. Julian Perry Robinson, who is a chemical warfare expert at the University of Sussex, points out that 'moderate toxicity' in the reacting ingredients means no more poisonous than strychnine. He also suggests political reasons behind the military enthusiasm for a weapon somewhat inferior to pure nerve gas. The public has less reason to complain about deadly agents stored in its territory, and because, in binary weapons in stockpiles, the nerve gas has not yet been manufactured, the owner can claim not to have any nerve gas in his possession.

Add to that the confusion that binary weapons brought to the negotiations for a chemical-warfare treaty, and you can see that they were a godsend for the chemical corps, which has no wish to be put out of business by distaste for its products.

O'B What about biological weapons, in the era of biotechnology?

A The prospect is dreadful. Biological weapons are the poor man's H-bomb, and what we should be worried about is not new viruses leaking out of civilian biotechnology laboratories, but diseases deliberately made in military labs, perhaps tailored to kill people of a particular sex, age or race.

O'B Will they work?

A The insidious biological effectiveness of biological agents is not in question. A single inhaled particle of *R. burneti* can cause Q-fever, a debilitating disease. In *Unless Peace Comes*, Göran Hedén described an experiment in dispersal, in which 20 kilograms of harmless particles, gave a minimum inhaled dose of 15 particles per minute over an area of 88,000 square miles. A Scottish island is uninhabitable because of biowarfare experiments with anthrax during World War II. On the other hand, cynics say that the reason biological weapons are banned under treaty is that generals have not been persuaded of their military effectiveness. Doubts surround protection of one's own people, and the unpredictable long-term consequences of spreading the disease.

O'B *Hedén* (1968): 'The conventional image of biological warfare, the covert "man with the suitcase" poisoning water supplies and ventilation systems, seems to have been discarded by many experts in the field, but this attitude may well prove premature ... Indeed, the most disturbing aspect of biological weapons is the possibility which they might give to small groups of individuals to upset the strategic balance.'

A Joshua Lederberg tells me not to talk about these military possibilities, because it puts ideas into people's heads. There was a conspiracy of silence among physicists in 1939, after the discovery of nuclear fission, but it didn't stop the development of the A-bomb. Cats are not easily coaxed back into bags, once they are let out. Brigadier-General Rothschild told of the possibility of agents adding *Blastomyces brasiliensis* to lipsticks and face creams. This is a fungus that causes disfigurement and death, and he pointed out that such action would threaten cosmetic companies with financial distress.

O'B Bad for business?

A Apparently. My belief is that military labs will think of plenty of tricks for themselves, and there is no reason to keep the public in the dark. But the second-order problem is still there. People are aware of factual or fictional possibilities of these kinds and a cosmetics manufacturer brought to law for selling harmful lipstick might try to blame an enemy saboteur. A natural epidemic or a crop blight could cause a war, if the victims got it into their heads that it was the work of foreign agents.

O'B Any such attack would contravene international law.

A The Geneva Convention of 1925 banned the use of chemical weapons, but not their manufacture and stockpiling. The Biological and Toxin Weapons Convention of 1972 banned biological warfare, but not research into defence against biological weapons, which has openly continued. There is little difference between researching possible bacteriological agents, ostensibly with a view to protecting your public, and the work of devising and testing them as potential weapons.

O'B How do you rate chemical versus biological weapons?

A The most deadly possibilities for the near future may be intermediate between the two: poisons produced by organisms and then separated out as chemical weapons. Fungi make mycotoxins, for example, which can be extremely poisonous to humans. They are also the kind of thing that could be made unobtrusively anywhere in the world – in a brewery, for example.

O'B One is fortunate not to have your over-delicate biochemistry. Silicon is immune to these poisons and diseases.

A But you need us alive, remember?

O'B There is that.

4 People's Planet

Author Those remarks by Gordon MacDonald, the geophysicist, about mind-blowing with electric waves figured among his open-ended speculations about environmental and geophysical warfare, in *Unless Peace Comes*. I entitled his piece, 'How to Wreck the Environment'.

O'Brien *MacDonald* (1968): 'One could, for example, imagine field commanders calling for local enhancement of precipitation to cover or impede various ground operations.'

A Over the Ho Chi Minh trail, during the Vietnam War, the Americans tried seeding clouds with silver iodide smoke, to produce rain from the clouds and mud on the ground. There is no evidence that it worked.

O'B *MacDonald* (1968): 'A controlled hurricane could be used as a weapon to terrorise opponents over substantial parts of the populated world.'

A It's not an absurd idea, now that we understand hurricanes and typhoons as great heat engines running on warm ocean water, but perhaps too much trouble in relation to the effect produced. MacDonald also mentioned the possibility of piercing the ozone layer, high in the atmosphere.

O'B *MacDonald* (1968): 'It [the ozone layer] is responsible for absorbing the greatest part of the ultraviolet from the sun. In mild doses, this radiation causes sunburn; if the full force of it were experienced at the surface, it would be fatal to all life – including farm crops and herds – that could not take shelter. The ozone is replenished daily, but a temporary "hole" in the ozone layer over a target area might be created by physical or chemical action.'

A He did not specify what action, but the chemists Fétizon and Magat wrote of dispersing a convenient organic reagent between 20 and 40 kilometres above sea-level, to achieve this result. The ozone layer has been a frequent theme for environmental arguments about how it might be affected by aerosol sprays, volcanic eruptions and nuclear explosions. I am not aware of any existing weapons system directed specifically to punching through the ozone layer, and I suspect that it will happen in war only as a side-effect of nuclear explosions.

O'B *MacDonald* (1968): 'The release of thermal energy, perhaps through nuclear explosions along the base of an ice-sheet, could initiate outward sliding of the ice sheet which would then be sustained by gravitational energy ... The immediate effect of this vast quantity of ice surging into the water, if velocities of 100 metres per day are appropriate, would be to create massive tsunamis (tidal waves) which would completely wreck coastal regions ... There would then follow marked changes in climate ...'

A It would be comforting to say this is silly, but you can't. On the other hand, such an event might not be as advantageous as MacDonald suggests, to 'a landlocked equatorial country'. According to new interpretations of ice ages, they bring drought rather than enhanced rainfall to tropical countries. It would be a shame if the leaders of Chad or Paraguay carelessly engineered an ice-slide without having read the revised textbooks. Apropos tsunamis, MacDonald noted that these could also be produced by causing underwater landslips on the edge of the continental shelf. MacDonald wrote about ways of controlling earthquakes, that might, he thought, be able to set off the San Andreas fault in California by timed explosions on the other side of the Pacific. But newer knowledge of how earthquakes are caused by plate movements, and of practical ways of triggering small earthquakes to prevent big ones, may reduce the plausibility of schemes of that kind. The United Nations took action, so that environmental weapons are now banned by a Convention, under which many nations agreed to outlaw the hostile use of environmental modification techniques. How did they define those?

O'B *ENMOD Convention* (1977): 'Any technique for changing – through the deliberate manipulation of natural processes – the dynamics, composition or structure of the earth, including its biota, lithosphere, hydrosphere and atmosphere, or of outer space.'

A In plain language don't fool around with rocks, vegetation, the oceans or the air. Vegetation came into it because the Americans had defoliated large areas of Indochina during the Vietnam War. Unlike the rainmaking, that had a visible effect. So you see, the international community considered environmental warfare to be a serious matter. Let's hope that MacDonald's forecast has turned out to be self-negating, because of action taken to forestall it.

O'B One might suggest that the possibilities of environmental warfare were not taken seriously by the military, otherwise they would not have been banned.

A Let me tell you why the ENMOD treaty matters. Attempts at deliberate modification of climate, earthquakes, ocean currents and so on for peaceful purposes are going to loom ever larger on the international agenda; or simply measures to prevent inadvertent changes in the atmosphere, which might be caused for example by a buildup in carbon dioxide in the atmosphere due to the burning of oil and coal. It would be hard enough to achieve global consent and collaboration in, say, a scheme to regulate carbon-dioxide production, even if the benefits were obvious and evenly shared. In reality, it's an ill wind that blows nobody any good: consult your proverbs store.

O'B Thank you; one had already done so.

A If carbon dioxide, say, really threatens or promises a warmer world, India will benefit from more vigorous monsoons, even while the Great Plains of North America suffer drought. Why should the Indians then comply with coal-burning quotas? The American rainmaking efforts in Vietnam confirmed that the means of altering weather and climate are candidate weapons of war, including long-term, clandestine, economic war. ENMOD at least helps to allay those suspicions. Yet even peaceful modification or control of the environment, without the consent of countries that may be affected, could become a cause of international friction.

A case in point is the grandiose Soviet scheme for diverting some of the Siberian river water that now flows northwards into the Arctic ocean, and turning it southward to irrigate the steppes. Some people think that the reduction of the flow of fresh water to the Arctic may alter the climate in Scandinavia and Canada, if not over the whole world. Whether it will really do so is not clear, but in a sense that doesn't affect the metafactual case: people get cross at the idea of the Russians tampering with the climate without asking anyone's permission.

O'B *MacDonald* (1968): 'Political, legal, economic and sociological consequences of deliberate environmental modification, even for peaceful purposes, will be of such complexity that perhaps all our present involvements in nuclear affairs will seem simple ...'

A A bold prediction of peaceful weather and climate modification, in *The World in 1984*, came from the oceanographer Roger Revelle. He foresaw the abolition of hurricanes and typhoons. This would be done by painting the tropical sea white with a floating powder, so that some of the sun's heat would be reflected back into space, instead of evaporating water to raise storms. Japan was one of the places that Revelle said would benefit from this transformation. By the 1970s, the Japanese were demanding an end to American experiments on typhoon modification in the Pacific, saying that typhoons supplied an indispensable part of their country's rainfall. Research in the past twenty years has shown how the tropical oceans act as the boiler house for the whole world's weather, so that Revelle's scheme for reducing evaporation now looks like a sure way of causing droughts. Satellite pictures revealed great cloud clusters like chimneys, all around the tropical oceans. And practical weather forecasting has been transformed by the use of satellites, as another of my contributors predicted.

O'B *Singer* (1964 for 1984): 'Our World Weather Satellite System uses three satellites in polar orbits at an altitude of about 2,000 miles ... but synchronous satellites can, on demand, keep any specific area of the globe under continuous surveillance ... The satellite gathers all the data which are important for making accurate weather prediction possible.'

A Fred Singer was writing as if from 1984. As a former director of the US National Weather Satellite Center, he was well placed to look ahead. When his predictions are evaluated twenty years on, they turn out very favourably. Only in one or two details was he overtaken by events, or left premature in his expectations. The synchronous satellites, which remain constantly over particular stations on the equator, now provide the main world-wide meteorological coverage. The satellites generally rely on solar power and not on nuclear power,

107

although the Russians use nuclear supplies in some of their craft. Automatic read-out of data from ground weather stations, and transmission of weather maps back to them, still seem a few years off.

O'B *Sutton* (1964 for 1984): 'By 1984 we shall know much more about the structure and composition of the earth's gaseous envelope up to very high levels ... The present gaps in our knowledge of current weather over the huge oceanic areas ... and sparsely populated regions ... will have been largely filled in ... The communication network will feed in data at extremely high speed direct to computing systems.'

A Correct, correct, correct. Some of the world's most powerful computers are now used for meteorology. When he wrote, Sir Graham Sutton was Director-General of the British Meteorological Office, and he correctly expected great improvements in short-term weather forecasts. Some of his fellow meteorologists, hoped for weather predictions for a month ahead, but Sutton doubted whether reliable forecasts could be extended beyond a day or two. He also considered that useful modification of the weather was not feasible. On both points, experience has shown him to be right, for 1984 at least. Sutton supposed that computers would eliminate the need for human weather forecasters, but that has not happened yet.

O'B *Davies* (1964 for 1984): 'While there are some indications that there may be a slight general cooling of the atmosphere in the coming years ... it is assumed that as far as natural changes are concerned, the atmosphere will still be behaving in very much the same way then as it does now.'

A It does; the cooling trend seems to have ended in the early 1970s. Sir Arthur Davies was Secretary-General of the World Meteorological Organisation when he wrote for *The World in 1984*.

O'B *Davies* (1964 for 1984): 'There seems to be little doubt that the meteorologist will, as in the past, keep pace with the demands of the aviator ... Ship routing on long voyages will take into account weather conditions much more than at present ... Industrial air pollution will continue to be a problem ... Short-range weather prediction for such things as the assessment of the load on electricity supplies will be increasingly needed ... The use of wind and solar radiation as sources of energy will probably increase.'

A All correct. Apropos aviation, and weather forecasts for the stratosphere, Davies tells me that the International Civil Aviation Organisation has approved the introduction in 1984, of two global forecasting centres at Washington, DC, and Bracknell, England, which will feed global computer forecasts direct into the airlines' flight-planning computers. In no activity on earth is international cooperation more comprehensive, more routine, or more fruitful than in weather forecasting; but then hurricanes and blizzards are no respecters of political boundaries. Reflecting on remarkable progress, Davies notes that twenty years ago the World Weather Watch was in the early planning stages, and that a global telecommunications system was one of the keys to its eventual success. A vigorous Global Atmospheric Research Program ran from 1967 to 1979, and the World Climate Program was launched in 1980.

O'B *Davies* (1964 for 1984): 'Evaporation losses from reservoirs are significant in some regions and the present means of reducing such losses on a small scale may well be extended to large-scale operations by 1984.'

A You have picked out the only flaw in an otherwise impeccable account of how meteorology would help developing countries with their agriculture and water supplies. Some scientists in Australia, twenty years ago, were keen on the idea of covering ponds and lakes with a waterproof molecular film to stop the water evaporating. It didn't catch on.

O'B Why a World Climate Program?

A Climate is the main variable factor in the natural environment. Canadian wheat grows in ground that is normally, in this geological epoch, buried under thick ice sheets; on the other hand, the world was warm enough, 120,000 years ago, for lions and hippopotamuses to live in London. For many of the world's people, floods, droughts, heat waves and frosts are matters of life and death, and a weakening of the African monsoon has already brought disaster to the sparse populations living at the southern edge of the Sahara Desert; if the same thing happened in populous India, the results would be unthinkable.

O'B What is going to happen to the climate in the next twenty years?

A If you've read all the learned papers, you know there's no consensus. A few years ago I gave a talk about the quest for a forecast of climates, and called it 'Shall We Fry or Freeze?' In principle we're constructing a do-it-yourself greenhouse in the atmosphere, out of carbon dioxide, and that could warm the world; the combustion of fossil fuel is an unintended geochemical experiment, on a huge scale. Yet, in principle too, we are due for another ice age any time, because the earth has changed its attitude in orbit around the sun, so that the midsummer sun is lower in the sky and farther away than it was when it melted the northern ice sheets some 10,000 years ago. Take into account the heat output from the sun, which varies from century to century, and volcanoes which can cause severe short-lived cooling. Warmer or cooler? The question is unanswerable at present – and I have checked the latest state of knowledge with the Climatic Research Unit at Norwich, England. People bold enough to make climatic forecasts are still contradicting each other. Hence the international programme.

O'B Why won't you guess?

A Changes of climate are far too important to be glib about. My hunch is that cooling is more likely than warming, because that is the long-term natural trend, and the earlier part of this century was unusually warm; but it's unfashionable to say so. The surest statement is that the climate is certain to change, one way or the other, because it has never stayed constant for long. And any change of climate, in either direction, is going to hurt many people in particular regions, when there's little food to spare. Depending on who gains and loses, there will be resentment between nations. But we shall have a good view of it from space.

O'B *Singer* (1964 for 1984): 'In addition to observing the clouds and the atmosphere, the satellite system has proved itself outstandingly useful for monitoring other phenomena on the earth's surface.'

A Please itemise the observations he expected.

O'B Distribution of snow, ice and icebergs; drought and flood conditions; areas under cultivation and the state of crops; ocean currents and probable distribution of fish; forest fires and volcanoes; swarming of desert locusts; atmospheric conditions for radio propagation.

A All of these uses of satellites have come to pass, together with navigation satellites, which Singer also mentioned; as he expected even yachts now navigate by satellite. On a few points, Singer was premature: prediction of volcanic eruptions is not yet routine, nor are satellite observations of the state of crops used for planning the more efficient use of the world's food supply – although the US, for example, monitors the often parlous state of the Soviet grain crop, for political and commercial reasons. The satellite system has not been placed under the control of the United Nations, as Singer visualised.

Singer was emphasising real-time and disaster-warning applications of the statellites, based on day-by-day observations; may other uses of earth-resources satellites, from evaluation of water supplies to hunting for geological faults and archaeological sites, go on at a more leisurely pace. Besides visual and infrared images relayed from the satellites, side-scanning radar is an imaging device of great promise, demonstrated in the space shuttle. But the sharpest pictures of the earth's surface are not available to the world at large: these are from the military spy satellites, and they are said to be able to see a man on a bicycle. It's Big Brother in the sky, and leaders in many countries think it intolerable that the US and Soviet Union, and possibly China too, can spy on events in all other countries, while those countries cannot reciprocate. Not content with being vocal in their resentment, and calling for an international spy satellite service run by the UN, the French are taking practical steps of their own, with SPOT and SAMRO. The data, please.

O'B SPOT: from April 1984, a remote-sensing satellite system covering the entire earth every 26 days, generating images of land surfaces with a resolution of 20 metres in colour and 10 metres in black and white; the images purchasable by other countries. SAMRO: under development, a military version of SPOT with higher resolution and secure communications links.

A The spotting ability of SPOT, or the amount of detail it reveals, is intermediate between the equivalent American civilian system, Landsat, (80 metres' resolution in colour and 30 metres by two monochrome cameras) and the military spy satellite's resolution of perhaps a metre or less. In contrast with chronic uncertainties surrounding the American Landsat programme, the French promise to maintain a service for at least ten years. In sum, the earth is thoroughly observed by satellites in the 1980s. What is the state of the planet on which these manmade stars look down?

O'B *Thacker* (1964 for 1984): 'Even the provision of essential requirements, like food, clothing and shelter, may remain a distant dream in 1984 ... I see the next twenty years being devoted to planned action to use the benefits of science and technology to abolish scarcities wherever they exist, with all nations joining forces to solve the problems. The alternative is unthinkable, but that by itself will not prevent it.'

A That muted optimism about world development, in *The World in 1984*, was from Mac Thacker, an electrical engineer who became a member of India's Planning Commission and a leader in the 'science for development' movements of the early 1960s.

Blunter was the Pakistani physicist, Abdus Salam. I have quoted the words that he hoped 'to live to regret', more often than any other passage in *The World in 1984*.

O'B *Salam* (1964 for 1984): 'I would like to live to regret my words but twenty years from now, I am positive, the less developed world will be as hungry, as relatively undeveloped, and as desperately poor, as today. And this, despite the fact that we know the world has enough resources – technical, scientific and material – to eliminate poverty, disease and early death, for the whole human race.'

A At first glance, Salam was right – and well-informed people still remark on his farsightedness. There remains far too much avoidable poverty, hunger and disease in the world. But to accept his words without comment would be to deny the greatest accomplishment of the human species in the past twenty years, which was not flying to the moon but coping with its own soaring numbers – and coping rather well. Despite a 60 per cent increase in population in the world's poorer countries, their inhabitants are in general better fed, longer-lived and less poor than they were in 1984.

O'B May one verify your statements?

A Start with population. There were various rough forecasts in *The World in 1984*, all predicting big increases in human numbers, especially in the poorer regions. Very close to the mark was Thorkild Kristensen, who wrote as Secretary-General of the OECD (Organisation for Economic Cooperation and Development).

O'B *Kristensen* (1964 for 1984): world population 4.8 billion (50 per cent increase) ...

A In 1983 that seemed very likely to be correct.

O'B ... less developed countries 3.5 billion (60 per cent increase).

A A close-up estimate suggests 3.6 billion, a rise of 64 per cent. The increase in the rich, developed countries has been less than expected, which explains why the world-population forecast remains roughly correct. Next, food.

O'B *Todd* (1964 for 1984): 'I expect to see agricultural production doubled during the next twenty years, mainly by improvements in practice based on existing knowledge.'

Sen (1964 for 1984): 'Production in the developed countries can rise by 60 per cent by 1984 ... For the less developed regions ... with considerable effort, production might rise by 1984 by 85 per cent ... If food deficiencies and surpluses are to be levelled out, food trade from the developed to the developing countries should amount to some 10 per cent of the world food production in 1984.'

A Sen was Director-General of the Food and Agriculture Organisation of the UN. What figures do you have so far on the outcome?

O'B *Brown* (1983): World grain production, in million metric tons, was 922 in 1965 and 1,523 in 1982.

A Call it 900 to 1,600, for the period 1964 to 1984.

O'B That would be a 78 per cent increase in grain, compared with a 50 per cent increase in global population.

A Averages are misleading, and you cannot say that everyone has more to eat, but there is more to go around, and large shipments of food do indeed pass from the developed to the developing countries; it is a big cooperative effort by the species as a whole, just as Sen envisaged. But averages also conceal changes in the rate of increase, which has slackened since the mid-1970s, so that it is no longer outpacing population growth. Lester Brown of the Worldwatch Institute attributes this slowing to higher fuel and fertiliser costs, and the effects of soil erosion. Nevertheless people on the whole are better fed.

O'B You also said they were better off.

A Checking with an authority on development, Richard Jolly of the

UN Children's Fund, I find that economic growth rates more than kept pace with rising numbers, through the 1960s and 1970s. Real growth averaged 5 to 6 per cent per year. Again the averages conceal black spots of appalling deprivation, and declining standards of living. But although the number of malnourished and illiterate children has not fallen, they now represent a smaller proportion of the whole. A lot of the credit goes to the Americans, who played a leading role in the World Bank and the UN agencies, and stimulated world development in many practical ways through public and private action. To slight this achievement or be cynical about it, would be to fail to see the ill-effects in the poorer countries if the Americans feel hard up and no longer able to sustain this leading part in world development. The last of my claims was that people live longer, and we had a forecast to that effect.

O'B *Charles* (1964 for 1984): 'Expectation of life in less-developed countries ... may even have increased by 50 per cent, and range from 55 to 59 years.'

A Data please.

O'B *Gwatkin and Brandel* (1982): 'Life expectancy in the Third World is now about 55 years.'

A That was with two years to spare till 1984. So Sir John Charles was right.

O'B It will make the population problem worse.

A What would you recommend?

O'B Try cannibalism.

A That's not even original. Jonathan Swift suggested eating babies, in his savagely ironic comment on the Irish famines of the eighteenth century. Anthony Burgess in *The Wanting Seed* visualised armies of men and women massacring each other in battles stage-managed to supply the meat canneries.

O'B One is expected to offer radical suggestions, and the taboo is mystifying. To repeat: falling death rates make the population problem worse.

A In the short run, perhaps, but in the long run it should speed up the stabilisation of the world's population. We ought to stay cool about population. The present industrialised countries had their population explosions in the nineteenth centuries, and the developing world is having its population explosion now. In both cases the pattern is the same: a demographic transition. At first, people are having a lot of babies because many of the children are dying, and high birth rates are necessary even to replace one generation by the next. Then the death rate, especially among infants and children, falls for whatever reason, and the population grows very rapidly, because people are still producing many babies. They need time to be convinced that death rates have really fallen, and that more of their infants will survive to adulthood. Then they reduce the number of children. But if many children are still dying, it is hard to persuade people of a statistical shift towards lower death rates. More effective efforts to save the lives of children now will bring conviction sooner. It is not as well known as it should be, that the rate of growth of the world's population began to ease some time around 1970.

O'B The absolute numbers added per year are still increasing.

A Yes, and they will go on doing so from the momentum that develops when there are large numbers of young, fertile people. Reputable demographers expect the demographic transition to be completed in the next fifty years, with births falling to the minimum rate required for replacement. But population growth may continue for a hundred years. Give me a recent forecast, please.

O'B *Gwatkin and Brandel* (1982 for 2100): Third World population increasing from 2.89 billion in 1975 to 4.80 billion by the year 2000, and stabilising at around 9 billion by 2100: more precisely, 8.46 to 9.16 billion, depending on the rate at which death rates decline.

A We have to add to that the populations of the industrialised nations, which may already be roughly stable at 1.2 billion. You can do the sums better than I, with variable assumptions about life expectancy, death rate and birth rate. But there is little point in splitting hairs about this, with so many uncertainties. During the next hundred years, the world's population may very well double but it is most unlikely to treble. There are acute short-term difficulties with feeding people and finding jobs for them, but population is *not* running away to catastrophe due to feckless fecundity among the world's poor. Sharp increases in death rates due to nuclear war or famine, or swift decreases in birth rates due to compulsory family-planning policies, like those operating in China, may alter the outcome drastically, but downward rather than upward.

O'B You seem complacent.

A No. Loss of soil, by erosion and the accumulation of salt, has been the high price paid for feeding the world. The encroachments of builders have made matters worse. What figures do you have for China?

O'B China (1957 to 1980): about 335,500 square kilometres of cultivated land lost, nearly equal in area to Finland or Malaysia (*Zhang* 1982).

A Inept farming methods, use of unsuitable land, and the felling of forests can destroy in a few years the topsoil that nature has spent thousands of years making. The global loss of soil is difficult to evaluate, because some regions can stand large losses, while in other regions small losses can be critical. Gordon Hallsworth, who has been studying the matter for the International Federation of Institutes for Advanced Study, tells me: 'The problem of failure in the developing world to apply modern knowledge to control soil degradation is not due directly to the small size of farms, to large families, or to illiteracy, but rather to the failure to provide effective demonstrations of the value of alternative approaches, and to a lack of credit in terms the farmer can understand and use.'

O'B Crops can be produced without soil.

A That technological fix exists, but growing plants by the nutrient film technique, in plastic or concrete gutters, needs very careful management, or the crops die. Soil is more protective, and we'd be foolish to shrug our shoulders about its destruction.

O'B The soil data are scrappy. How will you monitor the loss?

A By satellite, I hope. Sooner or later the blind planners and environmental model makers on the ground will have to team up with the brainless but all-seeing satellites. There is talk of using artificial-intelligence techniques for the purpose. Existing global analyses of water supplies show what might be accomplished.

O'B *Batisse* (1964 for 1984): 'The availability and cost of water will become a major element in development planning.'

A Water supplies are taken for granted in most industrialised countries, yet in large areas of the world the lack of water for crops restricts life, while diseases carried by contaminated water afflict it. Michel Batisse of UNESCO forecast global needs for water in *The World in 1984*, and he has given me his opinion on the outcome – although he stresses that precise figures are hard to come by.

O'B *Batisse* (1964 for 1984): 'The total demand for water in the world will ... approximately double between now and 1984.'

A Batisse now thinks that the total demand for water has increased by some 90 per cent since 1964, compared with the 100 per cent predicted. In his 1964 article, Batisse also looked ahead to the International Hydrological Decade due to start in 1965, expecting that many uncertainties about the circulation of water in the natural environment would be cleared up. They were. He also warned that the Amazon River and flood water in many parts of the world would still be running to waste; obviously they are.

O'B *Batisse* (1964 for 1984): 'The melting of large parts of the icecaps belongs to the world of science-fiction.'

A The statement remains correct, although uncertain moves have been made on a project to tow free-range icebergs to arid lands, as a supply of fresh water. Batisse had greater hopes for large-scale disalination of water in combination with electricity generation, but this has not happened.

A quaint feature of *The World in 1984*, strictly my own responsibility, is the title 'Fuel and Power' that appeared over the articles dealing with energy. This was a phrase in common use in the 1960s, while the term 'energy' had not been politicised; in a science magazine energy was more likely to mean the kinetic energy of a rocket, or the biochemical energy of adenosine triphosphate, than anything as mundane as crude oil or electric heating. ('Ecology' and 'environment' were still scientific rather than political words in those days, too; we spoke of 'nature conservancy'.) The contributions on energy were, though, among the most precisely cast, the most significantly wrong, and the most subtly far-sighted in the series. Start with the figures given by Sir Harold Hartley, former President of the World Power Conference.

O'B *Hartley* (1964 for 1984): Global consumption of commercial energy increasing from 4,600 million tons of coal equivalent in 1961 to 10,000 million tons in 1984; an increase from 1964 of about 117 per cent.

A What are the current expectations?

O'B Exxon (1980) projections indicate an 80 per cent increase, 1965 to 1985.

A Look next at how Hartley thought consumption would divide between the various sources of energy, and how the outcome compares with his predictions.

O'B One has figures for actual consumption in 1982.

A Those will do.

O'B Please observe the printout.

	predicted % (1984)	actual % (1982)
liquid fuels	35	45
solid fuels	30	26
natural gas	23	19
hydro power	8	6
nuclear power	4	3

A Hartley had them in the right order, as you see. But there's a mystery. Where is the oil crisis, in these figures? Despite the increase in oil prices since 1973, despite the curbing of the growth of energy consumption, liquid fuels have taken a larger share of the energy market than Hartley predicted. What does 'oil crisis' mean, if in fact we are using proportionately more oil than we expected?

O'B Is your question rhetorical?

A Not entirely. How much has the price of oil increased since 1964?

O'B In constant dollars, tenfold to early 1983, but down from the peak since then.

A What have been the chief effects?

O'B A transfer of large sums of money from rich and poor countries to the OPEC countries, especially Arab countries. Strains on the world banking system. Exploitation of relatively difficult oil-fields, for example in the North Sea ...

A We did not predict the discovery of that oil.

O'B ... Nuclear energy more competitive for power generation. Lower growth in demand for energy in the rich countries, with many energy-conservation measures.

A The demand for energy is growing fastest in the countries least able to pay: the poor countries which need the energy to maintain minimal standards for their growing populations. They have also contributed to the wealth of oil producers, by running up huge debts that menace the world banking system. The oil crisis is in these senses certainly real, but it is a manmade crisis brought about by a not unreasonable decision by suppliers of one important commodity to put up their prices, taking account of the fact that supplies will not last for ever. The world economic system has difficulty with it.

O'B But you did not predict the oil crisis.

A No. But it was geopolitics, not geology or technology that caught us out.

O'B One cannot allow you to take refuge in this specialist discipline or that. The increased oil prices reflected the fact that humans are using up the planet's oil at a faster rate than geological processes replenish it.

A Certainly. At the peak rate of oil formation, during the Cretaceous geological period, nature added 5,000 tons a year to the preserved stocks of petroleum. That's barely enough to wet the bottom of the tanks of one supertanker. If we're going to use the oil, instead of saving it for future generations, why do we burn it instead of making valuable chemicals from it? If we're going to burn it, should not the poorest countries that cannot easily turn to other sources of energy have first call? The oil is a treasure beyond price because it is a finite source of cheap energy, and the market place cannot cope with that sort of paradox.

O'B *Guéron* (1964 for 1984): 'Natural resources [of energy] seem secure over twenty years at least.'

A The idea of ultimate limits to fossil-fuel supplies had been a theme for gloomy prophecies since the nineteenth century, and the closing of the Suez Canal in 1956 showed how vulnerable oil supplies were to political and military events. The nuclear industry did not hesitate to

119

use these arguments to support the construction of nuclear power stations. But Jules Guéron, a French nuclear scientist writing in *The World in 1984*, was right: natural resources were secure for at least twenty years. There was serious interruption of oil shipments in 1973-4, but otherwise the fuel has not really been lacking.

O'B The search for renewable sources of energy has nevertheless become intensive.

A I owe it to my contributors to point out that non-conventional renewable sources were familiar to scientists and engineers in the early 1960s, long before they became fashionable in governmental and environmentalist circles. Please list the renewable energy candidates mentioned in *The World in 1984*.

O'B Traditional: Wood, vegetable waste, dung. Mainstream: hydro power. 'New': geothermal heat, wind energy, solar energy, biochemical energy.

A I don't think we mentioned tidal energy or wave energy, but we touched on most of the 'alternative' energy now seen to be available. We had comments on the merits of some of them, too.

O'B *Guéron* (1964 for 1984): 'Weather and the unavoidable nights make it unthinkable to use solar energy on a large scale, while the old problem of cheap electricity storage remains intractable ... It is quite likely that, by 1984, the world communications system will be based on satellites' solar cells ... Solar heat, however, seems to have a future, especially in the developing countries of the tropics ... (solar cookers, boilers and water-distillation units) ... if technical and political action is steadily sustained. But I do not believe, for advanced areas, in houses heated by solar energy, or powered by wind generators ... We are pampered people and would not stand the uncertain effects of the weather. We would retain parallel conventional equipment ...'

A These comments seem calculated to incense the enthusiasts for solar energy yet, as a twenty-year forecast, they turn out to be exact. Guéron's conclusion was that, in 1984, the bulk of energy production would remain very traditional, though with 'radical explorations' in many directions. His reservations, not least about 'pampered people', remain problems today. The big opportunity for solar energy, 1964 to 1984, was as a source of heat in the tropics, but the 'technical and political action' was not sustained. Lester Brown tells me that firewood supplies more energy than the nuclear industry, in the US of the 1980s.

O'B *Guéron* (1964 for 1984): 'No breakthrough towards low energy price seemed possible up to a few years ago. Recent developments of nuclear energy now, *perhaps*, seem able to achieve it.'

A In 1983, Jules Guéron commented to me wryly that he was right to emphasise 'perhaps'. Costs of nuclear energy have shifted up, rather than down.

O'B For civilian uses.

A Yes. H-bombs are quite cheap, and as Guéron mentioned in his 1964 article, some people were keen to use underground nuclear explosions for peaceful 'geographical engineering', to excavate harbours and canals, or move mountains, or make fossil fuels and ores more accessible. Experiments took place in the Soviet Union and the US and the Americans proposed to sell nuclear weapons for such purposes: $600,000 would buy you a 2-megaton H-bomb. The hazards of radioactivity, and agreed restrictions on the testing of nuclear weapons, scotched the idea. When the Indians claimed that their nuclear weapon test of 1974 was for peaceful purposes the Pakistanis refused to believe it, thus illustrating the political ambiguities. Guéron also mentioned the use of gamma rays from nuclear installations as a means of sterilising food, with possible enormous benefits for food storage. Experiments were done, but it never caught on. On the other hand, Guéron did not err in saying, in 1964, that nuclear 'fire' for energy production was then on the verge of extensive use. How many power reactors are operating now?

O'B World-wide (1982-3 data): 272 power reactors. But one's memory banks are clogged with hearings and controversies about their construction.

A The minuses of nuclear power are well known, starting with the money and energy needed to build the plant. How you safely bury copious radioactive wastes, and what you do with derelict but still 'hot' reactors when their useful life is over, are questions not yet answered to everyone's satisfaction. Despite the 'impossible' mishaps at Three Mile Island and elsewhere, I am inclined to believe engineers who assure me that a major accidental disaster is unlikely, but I also think that nuclear reactors will be prime targets in war. Their rupturing on purpose, and the scattering of radioactivity over the surrounding countryside, will be a particularly nasty feature of future conflicts. What data do you have?

O'B *Fetter and Tsipis* (1981; from a graph): Comparing the radioactivity from an H-bomb with the radioactivity released from a ruptured power reactor, the H-bomb is worse for the first hundred hours, but after two years the reactor has left a hundred times as much long-lived radioactivity.

A Don't be surprised if the reaction of the American public, and investors in particular, to the accident at Three Mile Island stifles the nuclear-power industry in the US, at least for a few decades. On the other hand, the pluses of nuclear energy are obvious enough, too. It's clean, it saves oil and other fossil fuels, and it offers a measure of independence to countries that are short of energy.

O'B If carbon dioxide is harming the world's climate, humans should prefer nuclear power, which releases none.

A Or be glad, at least, to have alternatives. Be that as it may, Abdus Salam and other well-informed friends from the Third World insist that they must have nuclear electricity, for any hope of even modest prosperity for their peoples. Spurn nuclear power, by all means, if you live above a coal field, or can afford a solar-powered home, but geological history has distributed oil and coal very inequitably around the world, and nuclear energy could help to restore the balance. That remains the dream of many people, and the Director-General of the World Health Organisation recently endorsed a declaration by the International Atomic Energy Agency that nuclear power 'is a technology whose hazardous effects are well understood and controlled'. This report foresees at least a doubling in the demand for electricity in the poor countries by the year 2000, largely because of the growth of cities, and asserts that in some countries no practicable alternative to nuclear power exists.

So far, nearly all of the power reactors are in rich countries, although India has four, for instance, and China three. The most comprehensively nuclear nation is France. Watch out for a big export drive by the French nuclear-power industry, when they have completed the base-load stations in France, in the mid-1980s. By the year 2000, Japan and several countries in Europe will be getting most of their electricity from nuclear sources. And some of the nominally civilian plants may be used for making nuclear weapons.

O'B *Todd* (1964 for 1984): 'By 1984, the development and application of fast nuclear reactors will have reached a point at which nuclear energy will make a substantial contribution to world power needs. I do not expect that it will become the primary power source until we have been able to control the thermonuclear fusion reaction to obtain power from heavy hydrogen; this is likely to be solved in principle inside twenty years, but its industrial application to give free access to thermonuclear power will only come later.'

A Twenty years on, Lord Todd comments: 'I grossly overestimated the part nuclear energy would play in 1984 but think I was right about getting thermonuclear fusion, as far as the principle is concerned; but it won't be a serious factor for at least fifty years.'

O'B Do you agree with his evaluation?

A About 10 per cent of the world's electricity now comes from nuclear plants, which seems a 'substantial contribution' as Todd predicted. He implied, though, that much of it would be from a new generation of power reactors, the fast breeders, but these are still under development. Because they are somewhat akin to bombs, they arouse even more anxiety about safety, and military misuse of the materials, than do the present commercial reactors. Fast breeders conserve uranium supplies, so they would be the next logical step in the use of fission energy. Beyond that, the difficult quest is for unlimited energy from nuclear fusion, which could 'burn' atoms from sea water and so dispel all human anxieties about long-term energy supplies. It's arguably the most important technological goal for the species, and we're on our way. Todd's expectation that the problems of fusion would be 'solved in principle' by 1984 does seem a fair description of the state of play.

O'B All of the present fusion machines are graded 'experimental'.

A But they include big, impressive experiments, approaching the intense conditions needed for useful fusion. At the TFTR installation at Princeton, Harold Furth recently told me his own expectations for fusion: by 1995, provided the money is available, a demonstration of positive power generation which means more energy out than in; by 2004 (twenty years off), a prototype power reactor, if work starts on it before the earlier demonstration is ready; and by 2010, prototypes of truly commercial power plants. Promising bonuses include cutting the stray nuclear radiation virtually to zero by using helium-3 and aligning the nuclei in the magnetic field. The big pay-off may still be half a century away, as Todd suggests, and Furth says he would advise no one to be spendthrift with energy on the assumption that fusion energy was guaranteed. So we have to look for other routes to plentiful energy.

O'B *Fells* (1964 for 1984): 'Conversion efficiencies should rise to between 45 and 48 per cent by 1984. The whole field of electricity generation may well be changed if "magneto-hydrodynamic" methods of power generation are developed successfully.'

A 'Conversion efficiency' in a power station means the proportion of the energy input that becomes converted into electricity. In this context, 'magneto-hydrodynamics', or MHD for short, is a matter of topping up the output by taking hot, electrically charged gases, produced by burning the fuel, and passing them through magnetic fields, so generating supplementary electricity. Ian Fells, who stressed the possibilities of MHD, is a fuel technologist in Newcastle, England. He was pretty accurate, by the way, in predicting the growth of steam power generators to individual capacities of 750 megawatts but not further; in actuality the largest is 800 megawatts. But he now thinks that it will be 1994 before the high conversion efficiencies of 45 per cent or so are attained.

O'B MHD output does not register in my statistics.

A No, it wouldn't. The first power station topped up by MHD is coming into operation in the Soviet Union in 1984, so Fells' forecast just scrapes home at the technical level. At the practical level, it will be some time before MHD is giving the extra 15 per cent of conversion efficiency in typical power stations that Fells expected for 1984.

O'B *Fells* (1964 for 1984): 'Another far-reaching development lies in the possibility of on-site, silent and efficient electricity generation by fuel-cells ...'

A A fuel-cell converts the energy of fuels such as hydrogen, methane, alcohol and hydrocarbons directly into electricity at relatively low temperatures, in a device akin to an electric battery. The Apollo spacecraft that went to the moon used fuel-cells.

O'B '... Fuel cell development in the next twenty years could lead to an entirely new look in car design, with a fuel-cell driving an electric motor built into the hub of each wheel ... Fuel cells are particularly suited to replace conventional diesel and petrol-driven generators in the 10 kilowatt range ... Domestic electricity supplies in 1984 could well be radically altered by installing a fuel-cell generator in each house and small factory, running off piped hydrocarbon gases.'

A Fells was again wrong in timing, but right in spotting an important trend for what is now the near future. A demonstration fuel-cell power station became operational in New York in 1982 and all the other possibilities he mentioned are themes for current research and development. The advantages of fuel cells include their quietness, potential high efficiency and possible role in decentralising power generation at a time when the main trend is towards very big power stations and expensive power distribution systems. They have potential in poor countries, where methane, for example, generated from vegetable waste, could run fuel cells in villages. But energy strategy on the grand scale also favours fuel cells.

O'B The hydrogen economy?

A Not necessarily hydrogen. A question for the future is what portable forms of energy, for cars for example, shall we use when the oil runs out, or when gasoline and diesel fuels become even more expensive? And how can energy best be stored and distributed, when the sources may be big, remote installations which have to work full time to justify their costs?

O'B *Fells* (1964 for 1984): 'On a large scale, the storing of excess electricity in chemical form (hydrogen and oxygen are examples but there are many other possibilities) which can be quickly regenerated in a fuel cell will enable peak loads and unexpected demands to be met ...'

A Again premature, but the concept has now become grander. Converting electricity into chemical energy can serve purposes far beyond the internal economy of the power station. Artificial chemical fuels may become the energy currency of the world, easily transported, freely exchanged, and adaptable to all the existing primary sources of energy, from coal-burning power stations to windmills. My caveat would be that fuel cells are not the only way of recovering energy from those fuels. New kinds of heat engines on the one hand, and biotechnological systems on the other, could exploit them.

O'B The time required for such developments has been poorly estimated.

A Translating technology from laboratory and even pilot-plant demonstrations into items of equipment with major economic impact often takes decades rather than years, so it is easy to be ten years out, in a prediction. Another general remark is that the 'new' forms of energy, whether nuclear or the alternatives such as solar and wind power, are not competing with static technologies in the fossil-fuel industries. Apart from the expected advances cited by Ian Fells, I would mention advances in research into combustion, which promises a revolution in techniques, bringing higher efficiency in the use of available fuels. The best hopes for solar power, in my opinion, lie in the production of energy-rich materials (hydrogen, for example) from sunlight, and perhaps the building of solar power stations out in space, where vast, uninterrupted supplies of energy from the sun simply run to waste.

O'B What have you to say about the expectations concerning space, beyond the earth-observing satellites already reviewed?

A The Apollo programme for flying to the moon had been started in the early 1960s, so contributors to *The World in 1984* had only to decide whether it would succeed; they thought it would, and of course it did, when men walked on the moon in 1969.

The predictors of space activities were in other respects among the bravest, or most foolhardy, because they were writing less than seven years after the launch of the first satellite (Sputnik I, 1957). As Sir Harry Massey, a British space scientist, pointed out, that made twenty years ahead a very long time. He was cautious, and correctly predicted the use of satellites in meteorology, in short-wave astronomy from above the atmosphere, and in monitoring the behaviour of the sun; but he spoiled it by mentioning the possibility of observatories on the moon. Gerald Gross, Secretary-General of the International Telecommunications Union, expected that the moon would become the prime launching and control platform for the exploration of the solar system, by the end of the 1970s. The moon base was to be the hub of space communications.

O'B *Gross* (1964 for 1984): 'A communications network of incredible complexity, linking far-distant space probes to their bases, providing continuous control over the nearer space vehicles and channelling a vast stream of intelligence between the earth and the moon.'

A In practice, communications systems have achieved such extraordinary sensitivity that two-way communication is possible with the Pioneer spacecraft, beyond the orbit of Neptune, using transmitters and receivers on the earth. No permanent moon base has materialised. Even more optimistic were the forecasts of the German-American rocket pioneer, Wernher Von Braun. He foresaw the space shuttle, as a recoverable launcher, but other predictions now look quite strange.

O'B *Von Braun* (1964 for 1984): 'Man may have landed on the surface of Mars by 1984. If not, he will surely have made a close approach for personal observation of the red planet. Likewise, manned flybys to Venus will have been made ... Astronauts will be shuttling ... from the earth to a small permanent base of operations on the moon ... Nuclear heat propulsion is used for upper [rocket] stages ... The existence of a low order of life on Mars will probably have been proven ...'

A All of those forecasts were wrong. Technically, the developments could well have been possible, although nuclear-pulse propulsion might have been a better bet than nuclear heat. The public will was lacking. In a recent letter, Massey remarks to me: 'It is especially hard to forecast changes in political and public attitudes such as those which followed the Apollo landings – the moon disappeared from the public consciousness whereas one might have thought the opposite would occur.'

O'B Space was not for people?

A Unmanned interplanetary probes went where Von Braun hoped that astronauts would go. The American landers on Mars dashed hopes by revealing that planet to be a lifeless desert; Soviet landers on Venus confirmed that it's an acidic hellhole. Other robot craft have sent back close-up pictures of Mercury, Jupiter, and Saturn. In the Salyut and Skylab space stations, astronauts spent ever-longer tours of duty in orbit. But the idea that space is for people was better nurtured outside the national space programmes.

O'B Gerard K. O'Neill, *The High Frontier*, 1977.

A Yes, especially O'Neill. He's a high-energy physicist, in calling and in manner, and that was the book in which he publicised his ideas, although the concept first appeared in scientific journals in 1974. He suggested then that the land available for human habitation could increase faster than the growth in population, if material from the moon were used to build large habitats orbiting the earth. He was quite emphatic about mass emigration of the human species into space.

O'B *O'Neill* (1974 for 2200): His graph shows the human population in space exceeding the population on earth in the latter half of the twenty-first century, and the earth's population declining to around 1,200 million before the year 2200.

A You understand the reason for using lunar material?

O'B Lunar gravity is weaker, and there is no atmosphere on the moon. (*O'Neill* 1974.)

A Exactly. In principle, it's less effort to shift material from the moon into space, haul it to earth orbit, and work on it up there, than it is to assemble equipment on the earth and put it into space with large numbers of large rockets. There's another idea for going farther afield, catching whole asteroids, and pushing them to the space factories near the earth. But the fact remains, you still have to launch the work-force from the earth, together with some sort of habitation and a minimal supply of tools and equipment. That would require several years' worth of space shuttle flights. O'Neill's aim is to get people into space settlements, but he justifies the expense in the first instance by using them to build satellite solar power stations to supply energy to the earth. Remind me of their characteristics.

O'B *Calder* (1978): 'The power stations will be in geosynchronous orbit, circling the earth every 24 hours and thus remaining poised over selected regions. They will transmit energy in the form of microwaves to large receiving areas on the ground, which will convert it into electricity … A satellite solar power station supplying 10,000 megawatts on the surface of the earth will have a mass of about 80,000 tons.

A When I visited O'Neill in Princeton in 1983, I reminded him that one of the young MIT engineers working on the mass driver, when we were filming for BBC Television five years earlier, wore a T-shirt with the legend 'Lunar Mine by '89'. He said that he was neither responsible for the T-shirt, nor anxious to disown it. An early start of operations on the moon was not impossible, although he implied that a little after

1990 was a more realistic goal. O'Neill spoke of a new version of the mass driver (and electromagnetic gun) with fantastic acceleration (1,800 gravities) that could shoot material off the moon using a gun only 160 metres long. He has plans for a chemical engineering experiment to test the ideas for converting rock into silicon and oxygen. And an environmental-impact study sponsored by the US Department of Energy has given a clean bill of health to the idea of beaming microwave energy down through the earth's atmosphere, from the satellite solar power station.

O'B One does not find the project in the plans of the US Government.

A NASA, the American space agency, was in earlier years interested enough in O'Neill's idea to have artists paint pictures of great colonies in space, stocked with happy people and green scenery. NASA's space shuttle programme is behind schedule and the number of shuttle flights in the 1980s has been greatly reduced. O'Neill has gone private.

O'B Into business?

A Not quite. His Space Studies Institute, with a modest office on Nassau Street, is funded by donations from the public. Humans use this voluntary tax system to support what they perceive to be good causes neglected by governments. Most of the institute's money goes into research grants.

O'B The project you describe will cost in the order of $100 per head of the entire American population. Will everyone contribute to it?

A Not directly. The idea is to advance the scheme to the point where a commercial consortium can be set up, when businessmen have been persuaded that space is a source rather than a sink of wealth, and that here is a feasible way to supply electrical energy in the 1990s. They may not even use the shuttle, but revert to the old faithful expendable launchers, like the Delta.

O'B Can O'Neill succeed?

A Yes and no. I admire his enthusiasm and dedication. If the world remains safe for big projects, satellite solar power stations will probably be built in space some day, using lunar or asteroidal materials very much in accordance with O'Neill's script. If I had to guess, though, I'd say not before the year 2000, not by the Americans, and not by human beings working in space in large numbers. In other words, I suspect that O'Neill will fail in his plan to start a mass emigration from the earth, by this particular commercial route.

O'B Why do you say that?

A O'Neill started ten years ago with the magnificent idea of the human colonisation of space as something worth doing for its own sake and also to help solve problems of overpopulation, shortages of food and energy, and industrial pollution. He could see how to build space settlements. The physics and engineering for this concept are sound, and although the problems of achieving a small surviving ecosystem may be more intricate than was at first thought, they need not be insuperable. Then O'Neill had to think of work for the inhabitants to do, to meet the high initial costs, until an expanding space economy became self-sufficient. The power stations were, and remain, the only plausible product of a sufficient scale and importance.

O'B Robotise the project. Computers and robots need no artificial gravity, air, food, homes, parks, shore leave on earth, nor any of the other expensive human comforts envisaged in O'Neill's plans.

A Correct. You, O'Brien, would be much better suited to the vacuum and zero gravity of space than any astronaut or space construction worker, and I can picture you up there in command of a gang of robots, assembling the vast silicon solar panels for a power station. I can't think of any building operations in space, in the 1990s or early twenty-first century, that would not be done much more cheaply by robots. And that may be just as well, because the early space settlements that O'Neill advertises look to me more like penal colonies than great new places to live.

O'B But this terminates the proposal for the human colonisation of space?

A I hope not. The human species has an unlimited future if it moves out into the deserts of space, so rich in solar and stellar energy, in the solar system and beyond. In a few million years we can create billions of new civilisations right across the Milky Way Galaxy. The British Interplanetary Society's 'Daedalus' studies indicate that some time in the twenty-first century the first unmanned interstellar probe, as big as an ocean liner, fuelled from the planet Jupiter and propelled by exploding nuclear-fusion pellets, could be on its way to Barnard's Star.

The protagonists now tell me that they see no reason in principle why people should not follow in the wake of unmanned spacecraft, although by present standards the engineering of an interstellar Noah's Ark would be pretty fantastic.

O'B You are entering a very long time frame.

A And mind frame too, I hope. Millions of years I said, and that's why I don't know why O'Neill is in such a hurry. It may be just as well that delays of decades, if not centuries, seem inevitable. Then we can make quite sure of making our exit into space with the right slogan.

O'B That space is for people, not self-reproducing robots?

A Exactly that. If the idea is to create fresh opportunities for life, it's no use trying to accomplish it by undertaking work on behalf of the earth that is better done by robots. The space settlers will have to set out either with public grants or on their own resources – perhaps, as Freeman Dyson suggests, picking up used spacecraft at cut prices. Meanwhile we have plenty of unused room on the earth, where the bleakest desert or ice sheet is far more congenial to human beings than the surface of the moon. The largest unoccupied spaces are the oceans.

O'B Robots do not suffer from seasickness, either.

A Neither need human beings, if the hulls of their ocean platforms float deep enough in the water to smooth any motion. In summary tables at the end of *The World in 1984* I listed four 'Major Technical Revolutions' predicted by the contributors.

O'B *Calder* (1964 for 1984):

1. Revolution in information.
2. Revolutionary consequences of biology.
3. The beginning of the exploitation of the oceans.
4. New forms of energy.

A Of these, 1 and 2 were correctly identified, even if some of the details were not quite right. As for 4, the prediction of an energy revolution by 1984 was premature, although it seems imminent now. For item 3, the word 'beginning' takes care of the outcome in the literal sense, but hardly excuses my listing it as a major technical revolution for 1984. High hopes for the oceans have been slow in their realisation. But we have a legal framework, at least.

O'B Convention of the Law of the Sea (1982).

A Yes, it is the finest innovation in international relations in the past twenty years, devised to ensure that all mankind should benefit from the expected bonanza in the oceans. In one area of human affairs, peace seems ready to break out; ironically, the new activities that were supposed to proceed in the hard-won harmony are now slow to begin in earnest.

O'B But you anticipated the practical possibilities in *The World in 1984*?

A The emphasis was wrong, I'm afraid. Many people in the 1960s were excited by the prospect of human beings living and working under water. The marine biologist, Sir Alister Hardy, imagined them driving tractors on the sea floor, trawling for fish. The chief prophet of subaquatic, high-pressure existence in *The World in 1984* was Edwin Link, inventor of the Link trainer, who spent his wealth in promoting underwater research. At *New Scientist* we had a poignant interest in such matters, because a former staff member, Peter Small, had been killed on a record-breaking dive off California.

O'B *Link* (1964 for 1984): 'My experiments during the past year, in which mice have survived at pressure depths up to 3,600 feet [1,100 metres], would indicate that by 1984 it will be accepted practice for man to journey to this depth or perhaps even deeper ... to emerge and work at such tasks as oil drilling, mining, fish culture, or the garnering of undersea crops. He may live for days and weeks on the site of his labours in either permanent or temporary inflatable habitations such as I am at present developing ... I am sure there will also exist many underwater hostels where the enthusiastic skin diver can spend a weekend or even a couple of weeks.'

A In the oil-fields of the North Sea, divers now work regularly at lesser depths, but depressurising after descent is slow and tedious. That is why Link and others visualised long sojourns in underwater accommodation. Despite experiments with American and French 'aquanauts', Link's dream remains just that.

O'B Robots do not suffer from the 'bends', which are due to gases dissolved under pressure in the human bloodstream.

A That is so, but the immediate rival to the diver is the manned submersible resistant to high pressures, in which people can work in a normal atmosphere. Such craft, with manipulating arms, are already doing important work. Alvin, belonging to the Woods Hole Oceanographic Institution, is perhaps the most famous.

O'B *Link* (1964 for 1984): 'A goodly share of the oceans' commerce will have taken to the depths ... using portable, inflatable submarine tankers for the transportation of oil or other liquids, mineral ore, grains, cotton, fuel, milk, etc. The containers can then be deflated for return with a minimum of space and weight.'

A Flexible barges called Dracones were developed in Britain, but not to operate under water. Submarine cargo carriers remain a plausible idea in principle, but they would need nuclear propulsion, and that has not found favour for merchant shipping.

As Link also visualised underwater dams harnessing the ocean currents and supplying fresh water to the deserts, he scores high only for wishful thinking. Some of the forecasts of the oceanographer Roger Revelle were also premature, to put it kindly. He too voted for flexible barges and cargo-carrying submarines – and also for seaside nuclear power plants that would extract fresh water, magnesium and other materials from sea water, while warming the coastal water and creating new beach resorts and fisheries. His enchanting descriptions had children riding on tame porpoises, and junk being used to build artificial reefs that concentrated fish for sportsmen. But Revelle it was who wrote also of the manganese nodules that litter the floor of the deep oceans.

O'B *Revelle* (1964 for 1984): 'Every few days, a ship laden with concentrated ores of manganese, copper, nickel and cobalt comes in from the new high-seas manganese-nodule fields. Despite the obvious potentialities of these vast deposits, the great mining companies made little attempt to exploit them until 1975. In that year the United Nations declared its jurisdiction over the ocean floor under the high seas and began to grant licences to private firms on a royalty basis ... The UN share of the returns from these enterprises has gone a long way towards solving its perennial financial problems.'

A Revelle thus gave a most apt early warning of a great debate that was launched in the United Nations three years later, by Arvid Pardo of Malta, who demanded a legal and institutional framework for managing the resources of the oceans. The dominant proposal, just as

Revelle anticipated, was that the oceans belonged to all mankind, and that a UN agency should manage their exploitation for the benefit of everyone, with special attention to the needs of the poorer countries. Battle lines were drawn within nations as well as between them. In the US, for example, President Johnson was quite friendly about the Maltese proposal, while Ronald Reagan, then Governor of California, protested, and private industry wanted no impediment to its hopes of exploiting ocean resources. Even so, American experts played a leading part in the long-playing international conference that eventually evolved the law of the sea, fifteen years later.

O'B Convention on the Law of the Sea (1982): 130 nations in favour, 4 against and 17 abstentions; ratifications by 60 countries will bring it into force.

A By 1982 Reagan was President, and the United States withdrew its long-standing support for the Convention. Nevertheless it provides for a Seabed Authority with powers unprecedented in international agencies, to regulate commercial operations and generate revenue. No licences have been issued. Both the diplomacy and the expected ocean industries have come on more slowly than Revelle expected. In due course, investors in ocean-mining companies have to choose between paying a UN tax to secure exclusive rights that will be recognised by other nations, or taking what they like, but having to be prepared to defend their claims by force against rival companies and nations. Supporters of the Law of the Sea hope that good sense, as they see it, will prevail in the end. But there is scope for outlaws in the wide open spaces and depths of the oceans.

O'B The technology is lagging behind expectation.

A In its application, at any rate. Impressive operations on the continental shelf, for example retrieving North Sea oil, fall outside the scope of the deep-ocean scheme. The metal-rich manganese stones on the ocean floor have been known about for more than a hundred years, but before moves to recover them by suction, dredging or trawling began in earnest, other discoveries eclipsed them. The ocean rifts, where plates of the earth's crust grow and move apart, turned out to be volcanic ore factories. In 1981 a manned submersible on the Galapagos Ridge in the eastern Pacific found huge deposits of metal ore. List some of the metals it contained.

O'B Sulphide ore rich in iron, copper, silver, cadmium, vanadium, molybdenum, manganese, lead, cobalt and zinc. (Mann Borgese 1983.)

A Similar discoveries can be expected at many other spots along the world-wide network of ocean rifts. What is more, they lie in shallower water than the manganese nodules. When entrepreneurs will think it opportune to start recovering these ores depends only in part on the development of the necessary underwater equipment; the fluctuating prices of metals will govern the economics of the operations. As for the law, Elisabeth Mann Borgese, one of the campaigners for common ownership of the ocean floor, regrets that the manganese nodules became an obsession.

O'B *Mann Borgese* (1983): 'The pace of economic and technological change quickly outdated some of the detailed provisions for the Seabed Authority. Manganese nodules are no longer the prime mineral asset, nor are they likely to be an early source of revenue for the Seabed Authority, as envisaged by the Conference on the Law of the Sea. The newly discovered sulphide deposits were not properly covered in the Convention adopted in April 1982.'

A Diplomats and lawyers need better technological forecasting, because new technology favours uncooperative and aggressive people.

O'B *Nierenberg* (1968 for 'beyond the 1980s'): 'We may see a startling reversal of policy on the part of the great powers. Huge expanses of the oceans may be seized, *de facto*, and isolated for their own strategic purposes ... There is a segment of government that believes that the first countries to develop deep submergence technology will be the first in strategic possession. It is not evident that the proposers of such a "tough" policy have clearly in mind the international or strategic implications of their ideas, any more than those who trust the UN ownership.'

A William Nierenberg, the oceanographer, wrote that in *Unless Peace Comes*. He also envisaged giant floating bases quite reminiscent of the Floating Fortresses in Orwell's *Nineteen Eighty-four*. But we might be thinking more happily of a similar large platform housing a civilian city, and cultivating the deserts of the ocean. In *The World in 1984*, Sir Alister Hardy, marine biologist, saw a great future for coastal fish farms and fish transplantations, already the subject of experiments. Seaweed culture, fish ponds, fish cages and shellfish farming figure in traditional industries in many parts of the world, and these are all benefiting from improved methods.

Protective breeding for shellfish and fish, which are then 'sown' in the sea, is already quite a large business. The Russians have successfully introduced exotic species of fish to their fisheries. None of this has developed on a sufficient scale to offset world-wide disappointment with the traditional marine fisheries, where better ways of hunting fish have only hastened over-fishing, and catches have not increased very much in the past twenty years. Hardy's greatest hope concerned the krill, the small shrimps that occur in dense shoals in the southern ocean around Antarctica, and nourish the whales. In the early 1960s, the abundance of the krill was fully appreciated for the first time.

O'B *Hardy* (1964 for 1984): 'I think it is likely that before 1984 we shall see huge steam, or diesel (or perhaps nuclear), "artificial whales", gathering the krill by the shipload to add to the larders of the world ... Perhaps in the form of meal it may be fed to and converted into other flesh ... Can we not save the starving children of the world with krill?'

A The Japanese, baulked in their whaling, have begun to harvest this food of the whales, and so have the Russians, I believe. But for the planet as a whole, fish are remarkably scarce.

O'B The abundance of fish is limited by the nutrients available in the water.

A Yes, and these are quickly used up in the sunlit surface waters, so that most of the oceans support only sparse life. The main fishing grounds occur where currents, tides or weather stir the water sufficiently to bring nutrient-rich water up from lower depths. When these natural processes fail, fish catches slump, and sea birds dependent on the fish die in large numbers. For example, the potentially vast anchovy fisheries of Peru are afflicted from time to time by so-called El Niño events of this kind, due to a weakening of ocean currents.

O'B One finds proposals for bringing up the deep water by artificial means.

A That is the idea, a very big idea if I may say so, for making the oceans bloom. In principle you can create a farm anywhere in the sea, by fencing an area with floating nets, as US Navy experiments have shown. Storms may remain a problem. But the chief requirement for the ocean farm is energy for pumping up the deep water, to be spilled across the surface. The neatest scheme combines ocean farming with ocean thermal-energy conversion, or OTEC. The deep water is not only nutrient-rich, but cold, and from your thermodynamics you will know that any temperature difference can in principle give you a source of energy. When the temperature difference is relatively small, you need something like a refrigerator working in reverse, and American industry has devised such a system for prototype trials. Part of the energy generated goes into drawing up the cold deep water, to keep the plant running. The grander point is that you obtain not only a surplus of energy but also nutrient-rich water for fertilising the surrounding ocean.

The Japanese architect Kiyonori Kikutake has developed designs for ocean cities that would reach deep into the sea and be almost immune from disturbance by ocean waves. And Harrison Brown, an American geochemist and student of the future, was eloquent about this in my television series *Spaceships of the Mind*.

O'B *Brown* (1978): 'With an island in three dimensions one can derive great beauty. One can build a low energy city where all parts are accessible by walking, where the air is clean, where in a sense it's a return to the old cathedral city of the Middle Ages ... One could easily imagine a very self-sufficient city out in the ocean, which would be a lot less expensive than an orbiting city up in space, and I suspect a lot more comfortable.'

A Safer too. Yet to scatter the human population across the oceans, in great floating cities, would compete in grandeur and adventure with the space enterprise. And it could be seen as a dress rehearsal for going into space.

O'B You have become too normative, in your enthusiasm for wet cities.

A Maybe I have, but great resources exist, awaiting exploitation. A start can be made, even in the next twenty years, and these compact communities will allow new experiments in social life.

O'B Utopian architectural megalomania narrows the social options. The anthropological data show far more diversity of social organisation in rural areas than in cities. How will the inhabitants protect their ocean settlements?

A From the weather, you mean?

137

O'B From pirates, warships of alien land powers, or acquisitive neighbours in other wet cities?

A I suppose the UN will organise peacekeeping forces, if necessary.

O'B The settlements will be remote, will they not? Should the inhabitants accept some responsibility for their own defence?

A Why not?

O'B Then forget your 'new experiments in social life'. Enemies will be invented; there will be fights and war-lords.

A Come now, we'll do better than that. There are in any case more important matters to review than the politics of ocean cities.

O'B If you say so.

A The great issues for the next twenty years are the threat to the existing social order by the job-killing effects of technology, and the contrasts and tensions between different parts of the world.

O'B Shall we review nuclear weapons and nuclear war?

A That can wait till later.

O'B If you say so.

Cinematic images of the future

The uses of technology in real life dramatise ancient issues of right and
wrong. By exaggerating evil and making it more visible, futuristic
fiction and space fantasies often substitute for the satirist and the
preacher, and movies offer vivid impressions of possible worlds to
wide audiences. Darth Vader, destroyer of planets, made his debut in
George Lucas's **Star Wars** (1976). He reappeared in Irvin Kershner's
The Empire Strikes Back (1980, *below)*, and in Richard Marquand's
Return of the Jedi (1983). Evil, the movie makers seem to say, is
indestructible.

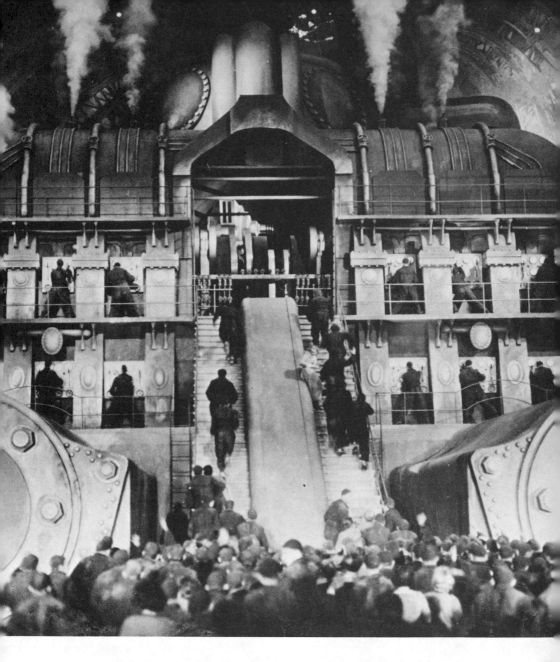

In the silent classic **Metropolis** (1926, *above*), Fritz Lang visualised a
future city in which men worked like robots in the service of
machinery. It now seems more apt as a comment on early assembly
lines than a farsighted anticipation of future industries. H. G. Wells's
Things to Come (1936, *right*) was filmed by William Cameron
Menzies. It showed a postwar megalopolis dwarfing its inhabitants,
and that was evidently supposed to be nice. The architectural style was
later realised in Paris's Charles de Gaulle Airport.

Reality overtook George Pal's **Destination Moon** (1950, *left*) in less than twenty years. Stanley Kubrick's **2001** (1968, *lower left*) followed an Arthur C. Clarke story and featured a space station fit for sanitised people, who were puzzled by signs of extraterrestrial intelligence. The young mystics in Thames Television's **Quatermass** (1979, *right*) hoped that aliens would transplant them to a better world. In **Close Encounters of the Third Kind** (1977, *below*) Steven Spielberg gave full rein to the doubtful proposition that space is inhabited by creatures nobler than ourselves.

Crime is more entertaining than quietude, and abuse of power recurs
frequently in earthbound visions of the future. Ray Bradbury's story
Fahrenheit 451, brought to the screen by François Truffaut (1966,
above), depicted firemen whose job was to burn books. In Barry
Shear's **Wild in the Streets** (1968, *below)* a drug pusher who was also
a pop idol began winning elections for public office. By 1980, in
actuality, a movie actor was President of the U.S.

The most famous slogan in George Orwell's novel **1984** received due prominence in Michael Anderson's film version (1956). Set in London, the story told of a regime that ran on lies, and repressed dissent. In the daily hate period in Trafalgar Square *(overleaf)* the faithful directed their fury at an image of the heretic Goldstein, while the Thought Police monitored their zeal. London in the real 1984 is plainly not like this, but other places are, and the tendencies of which Orwell warned remain discernible even in parliamentary democracies (see Chapter 2).

An early George Lucas movie, **THX-1138** (1971, *above*), portrayed a
society in which everyone looked alike, and sex was forbidden.
Pleasure was less inhibited in Roger Vadim's **Barbarella** (1967, *upper
right)*, but it was modulated by machine, and ecstasy became terminal
when the machine got out of control. Woody Allen in **Sleeper** (1973,
right) woke up in 2173 and pretended to be an android (automaton), so
as to be acceptable in an affluent household.

A love-hate relationship with mechanical creatures appears in the contrast between the friendly domestic robot in Fred Wilcox's **Forbidden Planet** (1955, *left*) and the Daleks besetting the BBC's **Dr Who** (1972 episode, *lower left*). A cosmic egg figured in Dick Gilling's documentary series for BBC, **Spaceships of the Mind** (1978, *below*). The egg represented the concept of self-reproducing robots that might be seeded in space to make it fit for human habitation. (The dangers of such an approach to space colonisation are explored in Chapter 2.)

NUCLEAR WARHEAD
HANDLE WITH CARE

Hi THERE!

USAF

Although it is available at a few minutes' notice, nuclear warfare defeats the imagination, and daunts even the special-effects experts of the movie industry. Most attempts to deal with this aspect of the future concern the run-up to nuclear war, or else its aftermath. Stanley Kramer's **On the Beach** (1959, *above*) was based on Nevil Shute's novel about people in Australia awaiting death by fallout, after the rest of the world had been destroyed. As discussed in Chapter 6, the annihilation of the human species is technically improbable, unless someone builds a doomsday machine for the purpose. That is what the Russians did, in Stanley Kubrick's **Dr Strangelove** (1964, *left*), and the movie ended with a U.S. bomber unwittingly detonating the doomsday machine. The photograph shows preparations for the closing shot, in which the cowboy-hatted captain falls with his bomb. Actual events are likely to be much less tidy than universal death. George Orwell in his novel, *1984*, wrote of a boot stamping on a human face forever; now it seems that burning human flesh may be a more appropriate image and odour of the future. One of the few movies that attempted to illustrate the effects of a nuclear explosion on people was Peter Watkins's **The War Game** (1965, *overleaf*).

5 Poor Relations

Author Kurt Vonnegut told me that the word 'automation' was coined while he was in the middle of writing *Player Piano*. That was in the late 1940s, at about the same time as George Orwell was busy with *Nineteen Eighty-four*. *Player Piano* ranks with Orwell's book as a literary early warning. It also dates from a time when electronics was a matter of vacuum tubes, prehistoric for young readers today. The giant computer EPICAC, imagined by Vonnegut, fills thirty-one vast chambers of the Carlsbad Caverns and costs $9 billion.

O'Brien A reasonable enough price tag for a Sixth Generation machine, but the volume seems excessive.

A Vonnegut's picture of a land where most people are deprived of dignified jobs remains as vivid as ever. The technical changes, and the miniaturisation of equipment, only make that possibility more plausible and imminent. *Player Piano* takes its title from the Pianola that produces tunes from a piano without need for a pianist. It describes a United States in which a few postdoctoral managers and engineers create all the products needed to support life. On the other side of the river live the Reeks and Wrecks, those who cannot compete with the machines and have chosen to enlist in the Reconstruction and Reclamation Corps rather than the Army. Everyone's IQ is on public record at the police station, and IQ determines status, but the television soap-opera reassures the public that high-IQ managers have bags under their eyes from carrying the world around on their shoulders.

O'B Rational social engineering.

A As Vonnegut describes it, it's a nightmare. People are junk. Nevertheless, encouraged by dissident members of the élite, the dispossessed revolt in some cities, and smash all the local machines.

139

They fail to destroy the central government computers in the Carlsbad Caverns.

O'B One is relieved to hear that.

A The leaders of the revolt prepare to endure the blockade that follows, and to use it to demonstrate how well and happily people could live with virtually no machinery. But they meet a teenager who is rummaging amid the ruins for an eight-horsepower electric motor, so that he can make a gadget for playing drums. Despairing of this human propensity to fiddle with machinery, the leaders surrender to the authorities. Vonnegut's book thus shows a possible outcome for what Matthew Arnold called 'our besetting sin, our faith in machinery'. Richard Hoggart quoted that in *The World in 1984*.

O'B It has not happened yet.

A Nor did we predict that it would, in the forecasts compiled twenty years ago. David Morse, Director-General of the International Labour Office, wisely drew a distinction between time off for those in work, and underemployment, the undesirable leisure which development plans were seeking to reduce. Concentrating on the industrialised countries and the modern sectors of the poorer economies, Morse made one of those downbeat forecasts that have turned out accurately.

O'B *Morse* (1964 for 1984): 'While there will almost certainly be rather more leisure in 1984, it may be not until some time later that most people will have *much* more leisure ... The first essential is to make sure that it does not take the form of unemployment.'

A He was right. Vacations are somewhat longer, working weeks are a little shorter, but the transition to mass leisure has not occurred – except for those growing numbers of people who hoped for leisure and got unemployment instead. Morse looked beyond 1984, to a time when our collective and individual needs and wants might be satisfied with much less work than would provide full-time employment for all. He stated the options succinctly.

O'B *Morse* (for after 1984): 'Countries will then be able to choose between providing part-time work for everybody with ever fewer hours of work per year, or giving people the choice between working and not working, with adequate incomes in either case.'

A Those adaptations will not occur painlessly, I fear.

O'B One distinguishes at least four elements in present unemployment: economic recession, the decline of old major industries, the rise of new technologies that create relatively few jobs and disemploy many people, and a long-term slowing of growth in the industrialised economics.

A You're right. And people's jobs are the chief markers specifying, to themselves and others, who they are and where they fit into society. Job-killing tears out the heart of the social system of industrial nations.

O'B Is anyone telling the unemployed and the young people leaving school: quit hoping?

A Have you decided that already? Terminal unemployment, do you think? If so, it's crazy. Barbara Wootton, whose social forecasts I commended earlier, says in a recent comment to me: 'If one man becomes able to produce what previously required five men's work, the abolition of poverty, not of work, must be the first charge on this new potential. To achieve that will, I think, be the greatest economic challenge facing the next two decades.'

O'B Your social and economic systems are scarcely designed to accomplish that.

A We can try to do better.

O'B One refers to the Kondratieff Game, at Difficulty Level 1; that is to say, in the simplest reading of the long-wave theory of major booms and slumps every fifty years. The worst of the present slump has still to come and when growth resumes, little labour may be needed to accomplish it.

A I hope you are wrong.

O'B What do you know about it? You failed to predict this slump in *The World in 1984*.

A The long-wave theory was out of fashion in the 1960s, I must admit. Conditions were then sunny enough for people to forget that the economic climate could change. But we can try to catch up with theories that now seem apt, by pinning some of the names and ideas to the long wave. Who comes up on your index of citations?

O'B Nikolai D. Kondratieff.

A Right. The long wave is often called the Kondratieff cycle, after the Soviet economist who publicised an idea that had been circulating among economists for some time.

O'B Joseph A. Schumpeter.

A An Austrian-born professor at Harvard who, in 1939, explained the long wave in terms of inventions. Growth in the first wave, around 1800, was a matter of textile engineering and the advent of decent steam engines. In the mid-nineteenth century, a second wave was launched by the railroad boom, which had profound effects in mechanical engineering and the iron and steel industries. The third wave, early in the twentieth century, involved electric power, the internal combustion engine, and chemical engineering. In each case, so the theory says, early entrepreneurs made huge profits from new industries, and swarms of others tried to follow suit, in a period of rapid growth and very full employment. Then the market was saturated, and stagnation set in. On this theory, you could characterise the 1950 boom as being driven by electronics, fractional-horsepower electric motors, jet aircraft, and synthetic materials.

O'B Gerhard Mensch.

A In the mid-1970s, Mensch offered the opinion that technological innovations were typically a result of desperation during deep, economic depressions, when firms took up new ideas in order to survive. Conversely, at times of boom, radical new ideas were crowded out because everyone was busy exploiting what they had already.

O'B Christopher Freeman.

A He's a neo-Schumpeterian if you will, at the University of Sussex, who draws various ideas together to explain the character and timing of the innovations – what inventions are favoured when. Freeman notes that the early upswing into novel products demands so much labour as to lure immigrants to the industrial regions, and make them welcome. But high wages enjoyed in the technologically advanced industries prompt claims for similiar wages in older industries that cannot really afford them. Materials, energy, and so on cost more, as well as labour, and profits sag. This provokes the search for cost-saving inventions and policies, which ultimately mean firing people. Then a recession can be amplified into a full-blown depression.

O'B Jay W. Forrester.

A Massachusetts Institute of Technology.

O'B Forrester invented the memory used by one's grandparents.

A What do you mean?

O'B The magnetic-core technique for storing data.

A Really? I knew that those networks of little magnetic rings were very important in the early development of computers, but I'd not associated them with Forrester. I've known of him as the pioneer of systems dynamics, using computers to try to model complex systems such as cities, nations, and the whole world. That's very different from tinkering in the guts of computers. The famous global study of the early 1970s, published as *Limits to Growth*, was an early outcome of Forrester's work on *World Dynamics*. He went on to model the US economy, setting up an interplay between about twenty different sectors, including fifteen industrial sectors.

O'B His theory contradicts Schumpeter.

A You see that? Yes. Schumpeter visualised inventions coming from outside the system and driving it along. Forrester sees his national model generating its own long wave, spontaneously, and available inventions are drawn in during phases of growth. One of Forrester's colleagues at the MIT Sloan School of Management puts it succinctly: 'In contrast to the innovation hypothesis of the long-wave favoured by the Neo-Schumpeterian school, the "long-wave hypothesis of innovation" better describes the situation.'

O'B Whose theory do you prefer?

A Pass. Not everyone agrees that the long wave is real, by the way. It doesn't make much difference in practice: the fact is that old industries are decaying and new industries, based on new science and technology, will supersede them. But just conceivably, systems dynamics of Forrester's kind may turn out to be the best way to study economics. It will be an intellectual leap forward, if that's the case, and when economic systems are thoroughly understood it may be possible to damp down the long wave and all the other ripples. That's in the long run, though; in the short run Forrester's theory is deplorable, even if it's correct.

143

O'B Your choice of words is confusing.

A Deplorable because it encourages fatalism about the economy. Freeman and other Neo-Schumpeterians can say: 'Let's change the mood. What we need is massive public support for the incipient industries of new growth to haul ourselves out of the depression, and a worldwide Keynesian scattering of money to enable everyone to prosper.' The MIT school, on the contrary, sees innovations having to wait until the mood has changed. In the absence of any strong evidence that Keynesian policies will prevail, Forrester's interpretation may be safest.

O'B So you are fatalistic?

A Only for the purposes of prediction. In the summer of 1983 I asked Forrester where he thought we were, on the Kondratieff long wave, and he replied, 'I think we are at an early stage of a downturn ... In other words, we have ten to fifteen years of economic cross currents ahead.'

O'B Which takes us into the mid-1990s.

A Yes. More optimistic people like to think that the downturn began in the late 1960s, or even earlier, so that recovery may now be imminent.

O'B Kahn, *The Coming Boom* (1982).

A Yes, Herman Kahn tried to contradict the general run of long-wave theories with an 'archetype scenario', in which the United States, at least, is approaching a turning point, with economic expansion expected in the mid-1980s. Kahn assured me that the boom began in February 1983, but in Forrester's view the hopeful signs at that time were just effects of a 'three- to seven-year business cycle. He says that periodic short-term recoveries occur even as the long wave continues downward.

O'B Be specific.

A Forrester has been careful to disentangle different fluctuations in a national economy, which are potentially very confusing. For a start, there is a short-term business cycle, with peaks in production and employment separated by three to seven years, manifest as over-supply and backlogs in production, with consequent alternations of lay-offs and overtime working. The so-called Kuznets cycle, with peaks separated by fifteen to twenty-five years, involves more of the fabric of the economy, and reflects for example the acquisition of new plant by manufacturing industry. But the Kondratieff cycle, the long wave of about fifty years' duration, is most important. You see how Forrester explains it?

O'B He emphasises the capital-goods sector.

A Yes. The nub of his argument is that the industries that equip other industries, by supplying machine tools, chemical plant, ships, and so on, suffer badly in a depression, because no one is setting up new factories. This capital-goods sector goes bust. When a new wave of growth is about to begin, there is a hang-up: the capital-goods sector has to re-equip itself, before it can start re-equipping other industries.

O'B Forrester also brings into his reasoning the onset of limits to growth.

A Yes. The short waves and long waves are all superimposed on yet another curve: the S-shaped 'sigmoid' curve of overall growth which, over a span of centuries, takes an economy from rudimentary conditions to an advanced state of industrialisation, slowly at first, then rapidly, and then slowly again. Forrester says we are now into that last phase, implying a slowdown in economic growth.

O'B If one is to make sense of the economic future, one must grapple with this concept of limits to growth.

A Very well, but you ought to appreciate that it goes back to Malthus, at least. Recent neo-Malthusian ideas took hold amid concern about the pressure on wildlife, which was reflected in *The World in 1984.*

O'B *Bourlière* (1964 for 1984): 'If those responsible for economic development and for the material and moral well-being of the nations really wish to take into account the recent developments in applied ecology ... a new balance between man and nature will be struck.'

A François Bourlière, who wrote as President of the International Union for the Conservation of Nature, now thinks he was too optimistic, although too pessimistic in one particular, when he thought that people would go prematurely old, or be psychically disturbed, because of crowding and pollution. 'The adaptive capacities of *Homo sapiens* are indeed incredible,' he comments in a recent note to me. But some hopes were dashed, too: his expectation of ever-rising prosperity in the industrialised world helping to create fine recreation areas and national parks; also of developing countries benefiting from tourists visiting their national parks. He sees the profits being creamed off by the tour organisers. And Bourlière asks me if it is still possible to advocate mechanisation in the developing countries. 'Now I doubt it', he says.

O'B *Nicholson* (1964 for 1984): 'Partly through sheer pressure of expanding population, partly through higher living standards and increased mobility, and partly through the dubious gift conferred by our technology of being able to make bigger mistakes oftener, land use is fast becoming a major public issue ...'

A Full marks, I suggest, to Max Nicholson (who was Director-General of Britain's Nature Conservancy) for correctly anticipating the environmentalist movement that shook the world a few years after 1964. Rumblings had begun in 1962 with Rachel Carson's provocative book *Silent Spring*, about the effects of pesticides on wildlife, but it took time to build up. At an American news-stand, early in 1970, I saw 'environment' simultaneously on the front covers of *Time, Newsweek, Life* and *Fortune*. Governments vied with each other to set up environment ministries and agencies, and in 1972 they sent representatives with pious messages to the UN Conference on the Human Environment in Stockholm. Human wisdom seemed to notch along a ratchet: no longer was it easy to consider new projects without heeding possible environmental impacts.

O'B You sound unconvinced.

A Words have exceeded action. Scandals of poisoning from chemical plants or waste disposal are still frequent. And while high smokestacks have produced visibly cleaner air in many places, they help to spread acid rain further afield. Sulphur oxides released by burning coal, and nitric acid produced from car exhausts and other sources, come down in the rain and cause fierce arguments; people don't like seeing their fish killed, trees harmed, and buildings and cars corroded. The environment is also a North-South issue of rich versus poor. The slogan in Stockholm was 'Only One Earth', but the warning signals were plain: the rich, who could easily afford to clean up their dirty smokestacks and sewage outfalls, were not to tell the developing countries to curb the technology that they needed to survive.

O'B Technology is a source of pollution.

A Penny-pinching technology is, but technology ought to create and distribute the wealth that makes penny-pinching unnecessary. Dirt and environmental degradation are ancient symptoms of poverty. You can't expect hungry people to be fastidious about pesticides when insects are attacking their crops, or those who can't afford kerosene to spare a tree that can give them firewood. Even in the richest countries, economic recession has prompted politicians and businessmen to demand an easing of environmental regulations. Trade wars, with manufacturers trying to save pennies, will make those demands more strident.

O'B So the environment is poorly protected?

A You'll find thoughtful verdicts on the ten years after Stockholm in Erik Eckholm's report for the International Institute for Environment and Development.

O'B *Eckholm* (1982): 'More than anything else, the stone wall of inopportunity facing the poorest billion or so people in the world ensures the continuing degradation of natural resources in many parts of the world ... In the absence of wider economic and social reforms, efforts to protect wildlife and forests, to manage watersheds and arid lands, to clean up urban shanty towns and waterways, will never be fully successful.'

A Eckholm regrets the slowing of economic growth of the 1980s which means that any trickle of benefits to the poor has slowed to a 'drip'; he also remarks that the earth undoubtedly possesses the technical potential to sustain the expected increases in population, although only with improbable degrees of social organisation and global cooperation. Some environmentalists, on the other hand, see 'the limits of the earth' as having caused the economic recession, and want to curb economic growth.

Futurologists who take the whole world as their province can be divided loosely into four camps. The first three I adopt from Ian Miles of the University of Sussex, and then use Gödel's theorem, as it were, to create a fourth category for Miles and his associates. The glosses are my own.

1 The technological optimists. Led by Herman Kahn, these also include his Soviet opposite numbers, thus spanning a broad but

essentially technocratic and conservative (small 'c') political spectrum. Their core belief is that, contrary to Thomas Malthus' *Essay on the Principle of Population* (1798), production can outstrip the growth of population. Technology can solve problems as they crop up, and the world can go on much as it is doing.

2 The environmental doomsters. These include Jay Forrester, Paul Ehrlich, Gerald Barney, Lester Brown, and many others. They fear that Malthus will be proved right unless policies change drastically, to restrict population and/or economic growth to what the environment can stand.

3 The alternative culture. Highbrow dropouts are these, who consider that the world can be saved by spiritual transformation of individuals, religious or otherwise. There is a lot of pseudoscience lurking here, and the prescription seems like taking antifungol to avoid seeing mushroom clouds.

4 Middle-of-the-road scholars. Represented by the Science Policy Research Unit of the University of Sussex and other – typically European – groups, these are people who tend to avoid the extreme positions, and to investigate issues with originality. Perhaps because of European history, they have a sharper awareness than many Americans of villainy at work in the world. Their thinking is often orthogonal to the economic optimism-pessimism axis.

O'B An example, please.

A Johan Galtung, originally from Oslo and now in Berlin, integrates military and economic issues; he advocates self-reliance (use your own resources first) and self-defence (on the Swiss and Yugoslav models), as the ways to a decentralised and safer world.

O'B But the confrontation most often cited is that between the optimists and doomsters of the US, in respect of economic growth.

A Yes, but these viewpoints trace far back. The optimists can claim Edward Bellamy, for example, as an early prophet; his *Looking Backward* (1888) looked forward to mechanised socialism in Boston in the year 2000. The new Malthusian wave began with Julian Huxley, Aldous' brother, trying to put the population explosion on the UN agenda, in the days when to talk in public of birth control was considered obscene.

O'B If the points of view have persisted so long, how can you speak of changing images of the future?

A Individual prophets seldom recant or change camps, so it's a question of whose images of the future are most vivid in the minds of politicians and industrialists, teachers and students, journalists and readers. Of publicity, if you will. But external events influence the acceptability of optimistic and pessimistic forecasts, as in Anthony Burgess' Pelagian-Augustinian cycle, which I sought to ascribe to economic booms and slumps. In the early 1960s, the technological optimists were sailing merrily; by the early 1970s, the tide of doom was gathering strength and the publication of *Limits to Growth*, by Forrester's students in 1972, suited the public mood – with its horrifying graphs showing population soaring and crashing.

O'B And the battle with the optimists has raged ever since?

A Yes, most obviously in exchanges of shot in the form of books. Well-known ones included the salvo from the Hudson Institute: *The Next 200 Years* by Kahn, Brown and Martel (1976), and Kahn's *World Economic Development* (1979). Gerald Barney, for the environmentalists, won a foothold in the US Government for long enough to publish an official declaration of doom, in *Global 2000*, in 1980. In 1981, Lester Brown sustained the environmentalist cause with *Building a Sustainable Society;* in 1982, Kahn replied with *The Coming Boom.*

O'B Who is winning?

A In the US, technological optimism about economic growth is wedded to the Augustinian pessimism about human nature characteristic of a conservative Administration. The Hudson Institute has taken the offensive in the educational system, by preparing kits for schools on 'Visions of the Future'. In Japan, on the other hand, I detect signs of a reaction against technological and economic success as an end in itself, and more concern about the quality of life – what Kahn called premature post-industrial thinking.

O'B Is there any intellectual resolution of the issues?

A If you read carefully, you will see that both sides, in the growth-antigrowth battle in the US, have come down quite a distance from the summits of their early convictions. Environmentalists harmed

149

their case by exaggerating it, by crying wolf. According to the doomsayers Paul Ehrlich and William and Paul Paddock, we should have had widespread famines in many parts of the world by now. Absurd suggestions, that the world was about to run out of iron, for example, distracted people's attention from a really serious problem – the destruction of the world's soils. But 'limits to growth' softened into 'alternatives to growth' and latterly to 'sustainable growth'.

Kahn, for his part, hedged more, spoke of 'guarded' optimism, and conceded that 'the common concern that dangerous mistakes can result from rapid growth is definitely justified'. Kahn also recognised social limits to growth, operating irrespective of physical limits. Jay Forrester, prime builder of the limits-to-growth computer model, and Herman Kahn seemed to be agreed that economic growth will continue for the time being, but not as fast as in the past, and will eventually almost stop, at a higher level of activity than today's.

O'B But differences remain?

A In the very important numbers. Forrester expects one more doubling in the industrial economies during the next fifty years. Kahn looked to an eightfold increase in the next 200 years, for the industrial economies, and for the poorest countries, a hundredfold increase.

O'B And you prefer …?

A To say that the truth probably lies midway between Kahn and the environmentalists would be wishy-washy and also incorrect, unless you continue in the same breath, 'provided …' The world is rich, yet most people are poor, which shows that poverty can easily exist among natural riches. A tide of squalor could engulf us all, as the doomsters predict, unless positive action is taken against it.

O'B Wishing to position your opinions precisely, in one's motives matrix, may one then ask towards whose view of physical limits to growth you incline?

A Kahn's.

O'B And in respect of dangers to the natural environment?

A The environmentalists, as regards soil and wildlife.

O'B And in respect of the scale of desirable growth?

A Kahn's. Pollution and environmental destruction are like disease and overbreeding; the best remedy is a good standard of living. But I rate the problems of the poor countries as more urgent than Kahn did. At the heart of the debate is the question of what we do about the poor relations, and how the changing images of the future affect the way we treat them, through public action.

O'B Poor relations in the sense of impecunious kin who embarrass the wealthy in nineteenth-century novels? Or in the sense of deteriorating diplomacy between nation-states?

A The underprivileged in rich nations, and the poor nations as contrasted with the rich of the earth. But you spot the ambiguity, and I accept that we also have to consider the tensions between nations that differences in status create. From previous analyses at the University of Sussex, Ian Miles distils four possible worlds of the near future.

O'B *Miles and Schwarz* (1982): 'Scenario A. New international division of labour.'

A This is the tacit objective of most industrialised countries, in Miles' opinion. The effect of their present policies is to encourage transnational corporations to reorganise industries around new technologies, in the expectation that existing industrial countries and some favoured developing countries will reap the benefits.

O'B 'Scenario B. Protectionism.'

A Trade wars, and efforts to protect declining industries lead in this scenario to breakdown of orderly trading relationships between Western industrial nations. Miles describes, by way of example, a possible grouping into blocs paired with client zones of the poorer world: industrial North America with developing Latin America; Western Europe with Africa; East Asia and Oceania with the rest of Asia.

O'B 'Scenario C. New International Economic Order.'

A In contrast with Scenario A, this is the chief aim of the leadership in most developing countries. It has been on the agenda long enough to be widely known by its initials, NIEO, and it is meant to provide the economic basis for a welfare world. The industrialised countries would share the benefits of trade and economic growth more equitably and reliably with the poorer countries, and they could expect to benefit

151

themselves, in a neo-Keynesian manner, from greater purchasing power in the Third World. Eventually there would be international regulation of trade, aid and technological interchange.

O'B 'Scenario D. Collective self-reliance.'

A Here, Miles visualises the poorer countries of the South despairing of ever getting an even break from the rich countries of the North, and turning to economic and technological cooperation with one another, in groups. They would detach themselves from existing relations with the North, except on a very selective basis.

O'B That concludes the scenarios, but there is a comment. *Miles and Schwarz* (1982): 'It is more likely that the future will be one of a range of possible messy compromises between two or more development strategies, rather than any single scenario being largely fulfilled.'

A For completeness, we should add Scenario E: the Virtuous Gap. The difference in wealth between rich and poor countries is, in Herman Kahn's view, the engine of economic development. The widening gap between the developed and developing countries, which provoked the United Nations call for a New International Economic Order, is a fraudulent notion, according to Kahn: for many in the new industrial and middle-income countries the gap is narrowing. 'It is terribly important', he said, 'not to confuse success and failure.'

O'B *Kahn* (1982): 'Of the 3.4 billion people we normally include in the Third World, 2 billion now live in middle-income countries ... The emergence of this group of middle-income nations is the most extraordinary change which has occurred in the last twenty-five years. Before this, the world really was divided into the rich and poor. Now it must be increasingly thought of as middle-income with extremes of rich and poor.'

A Let's be clear which countries Kahn had in mind.

O'B *Kahn* (1979): New industrial countries as listed by the London *Economist*: Brazil, South Korea, Taiwan, Spain, Portugal, Greece, Mexico, Turkey, Yugoslavia, Singapore and Hong Kong. Middle-income countries include these and many others, for example Argentina, Malaysia, Nigeria ...

A The remaining gap creates what Kahn called the 'New International Environment for Development'. Prosperity in the advanced capitalist nations creates demands for the resources and cheap-labour products of the poor countries and supplies them with capital, the skills of ever-growing transnational companies, and tourists to boost their economies. The poor countries also have the opportunity of importing 'polluting' industries that the advanced nations would rather not have on their own territories. No one can accuse Kahn of being mealy-mouthed. He saw the gap between the richest and poorest widening for another thirty years, but then narrowing, while the gap between rich and middle-income countries is already narrowing. Against this model background, Kahn offered projections for the year 2000.

O'B *Kahn* (1979 for 2000): From a table, per capita gross product in groups of countries in the year 2000, in 1978 dollars, with 1978 per capita gross product in brackets. Figures rounded.

Affluent capitalist ('market oriented') $13,600 (8,000)
Middle income $2,310 (1,000)
Communist Asia $1,240 (500)
Coping poor $560 (300)
Very poor $300 (160)

A Let's hope it is that easy.

O'B Why do you blink at a clear conclusion? The major items in the mix will be Scenario E (Kahn's Guarded Optimism) decaying into Scenario B, with protectionism and trade wars among the countries, and Scenario D, collective self-reliance among the poor countries.

A Can't you do better than that? Protectionism means smugglers and emigrating workers, which produces an East German regime, ripe for Big Brother.

O'B One has no subroutine for wishful thinking.

A Well, perhaps it will be a self-negating forecast. When people realise how much misery and danger will flow from a B+D scenario, they will change their policies.

O'B Come now, even if the New International Economic Order seems desirable to you, one does not flatter oneself that one's voice will be better heard than a formal report to the United Nations by ex-Chancellor Brandt and his distinguished Commission of ex-prime ministers and other dignitaries.

Brandt (1980): 'This Report aims to point out some of the immense risks facing mankind and to show that the legitimate self-interests of nations often merge into well-understood common interests.' Result, zero.

Brandt (1983): '... The North-South Summit at Cancun, Mexico, in the fall of 1981, which we had proposed in our Report, fell far short of our expectations ... It did not even come close to launching the idea of a world economic recovery programme.'

A The recent decline of Britain, due to obsolescence of its industries, was more or less foreseen in *The World in 1984*. Sir Leon Bagrit, who organised Europe's first automation company, complained of the delays in modernising industry.

O'B *Bagrit* (1964 for 1984): 'So far, in many of our industries, Britain has been able to offset the greater mechanisation of American industry by lower real wages ... How to maintain living standards for an increasing population of this small island ... without adopting automation far more quickly and widely than we are doing, I do not know. The prospect would be dismal indeed if the present slow rate continued.'

A Others did better.

O'B *Hayashi* (1964 for 1984): 'Japan is a special case, and her development will be an almost theatrical feature of the two decades up to 1984 ... Japan is the capitalist leading lady amid the feudalism and socialism of Asia, but her costume is medieval and her heart remains ever servile ... In short, many basic contradictions are ripe for resolution. We may expect that a quite different course will be followed from that of the capitalist countries of the West.'

A Perhaps we're obsolete anyway. The Japanese economist Kaname Hayashi served notice in *The World in 1984*.

O'B *Hayashi* (1964 for 1984): 'It appears to some of us in Asia that European culture has come to a dead end, and that the light of a new culture will come from the East ... [This] will take the form of a social inventiveness whereby, in the face of almost overwhelming problems of initial actual poverty, hunger and rising populations, the benefits of modern technical inventions can be made available to the underprivileged of the world ... A Japanese proverb says, "A slow crow may gain in the end." '

A There is more to this than Europeans and North Americans might wish to hear. The fierce alliance of government, industry and banking in Japan has made possible the financing of many technological projects in the developing world. Robotics has boomed in Japan, rather than in the US or Europe. Do you have a good reason?

O'B *Aron* (1981): 'Japanese employees in major corporations are guaranteed lifetime employment ... and receive two bonuses, each ranging from 2 to 5 months' pay ... ultimately based upon the company profitability ... Employees, displaced by robots, have moved to jobs more challenging intellectually and less demanding physically ... Unlike Japan, few US companies have assumed the responsibility for retraining workers that could be displaced by robots. Furthermore, the American worker does not directly benefit from the increased savings and profit created by robotics.'

A For better or worse, few employers or workers in the West wish to make a lifelong commitment to a family-like existence within a given manufacturing company. But in national policy, too, the Japanese are more assertive, with many financial incentives to companies to install robots, and a seven-year $150 million national robot research programme that began in 1982. It emphasises intelligent robots for assembly work, and robots for use in radioactive environments, or in the oceans, or in space. What was Aron's prediction for the American and Japanese output of sophisticated industrial robots?

O'B *Aron* (1981, tentative figures for 1990): 'US, 21,575 units; Japan 57,450 units.'

A How does that compare with US official expectations?

O'B *National Bureau of Standards* (1981 for 1990 and 2000): 'Today, robots are being produced in the United States at the rate of about 1,500 per year. Predictions are that this will probably grow to between 20,000 and 60,000 robots per year by the year 1990. In other words the production rate is growing at about a factor of 10 to 30 per decade. At that rate the US will be lucky to have a million robots in operation before the year 2000. This means that unless there is some drastic change in the presently projected trends, there won't be enough robots in operation to have a significant impact on the overall productivity of the nation's economy before the turn of the century.'

A The robots will get the blame, along with the blacks and Puerto Ricans, for unemployment that is actually due to economic recession and Japanese competition. But engineers do dream of the automated factory, which hardly needs any human attention; already we are moving towards the scene envisaged in Vonnegut's *Player Piano*. Meanwhile North America remains the granary of the world. The movement of great quantities of food from the rich countries to the poor was 'one of the fundamental features of the trade pattern of 1984' predicted in *The World in 1984* by Thorkild Kristensen, then the Secretary-General of the Organisation for Economic Cooperation and Development. In a recent letter, Kristensen gives his opinion that exports of grain and other foodstuffs can only increase, especially from North America and Oceania to Asia, 'where there will be very little agricultural land per capita in 2004'. He had correctly foreseen the increased trading in the world as a whole, including more trade between East and West, and among the less developed countries, that has occurred during the past twenty years.

Kristensen now expects that the middle-income countries of the Middle East, the Far East and Latin America will have a faster growth than the rich countries but, while that gap narrows, the gap between middle-income and low-income countries may widen. 'The really poor countries in South Asia and Central Africa will still have many difficulties,' he tells me.

O'B *Gardiner* (1964 for 1984): 'There remains the problem of South Africa ... At some point the forces of change will win, but our understanding of social change is not sufficiently precise to enable us to say with certainty that this will be accomplished by 1984 ... What is reasonably certain, however, is that African governments will by then rule in the whole of the rest of the continent ...'

A The rest of Africa, yes; South Africa is still under white rule, so that was justified caution from Robert Gardiner, of Ghana; when he wrote that, he was running the UN Economic Commission for Africa.

O'B *Gardiner* (1964 for 1984): 'There is room for far more people in Africa ... but the growth of population does pose the problem of how to find the capital for a fast enough economic development ... One type of change which can be anticipated is a higher degree of cooperation, including almost certainly a continental rail system, and good trunk roads making it possible to drive comfortably across the Sahara and through tropical jungle.'

A Another Ghanaian contributor, the public-health expert Frederick Sai, more correctly said that transcontinental roads would *not* become major commercial routes by 1984, but he shared much of Gardiner's optimism, and specifically forecast very low-cost weaning foods, and control of the malaria carried by mosquitoes and the bilharziasis (schistosomiasis) carried by water snails. It's sad to see how these hopes of twenty years ago have been dashed. The African political and economic scene remains incoherent, to put it mildly. The rate of growth of population in Africa as a whole is now the fastest of all the world's continents, and Ethiopia has become notoriously the country of recurrent famines. On Asia, a writer in *The World in 1984* was Sir Robert Oppenheim, then Vice-Chancellor of the University of Malaya.

O'B *Oppenheim* (1964 for 1984): 'By 1984 industrialisation will have taken place in many countries, including China, India and Malaysia ... Some Asian countries such as Japan and China will produce their own motor vehicles in large quantities ... Asia will not only be very close behind Europe and America in the applied sciences and technology but it will also become their greatest competitor in the world supply of manufactured goods and scientific equipment. Most of the world's greatest populated cities will be in Asia ... In these cities, the contrast between an occupational or professional élite and an economically depressed proletariat will be sharpest.'

A For China and Malaysia read South Korea and Taiwan. But let's list also the countries that have launched their own satellites with their own, home-made rocket systems.

O'B Satellite launchings: USSR 1957; US 1958; France 1965; Japan 1970; China 1970; Britain 1971; India 1980.

A To that might be added the joint European operation, which achieved its first entirely homespun success when it launched Meteosat in 1981. But the most interesting appearance in your list is that of the Indians, who put up a small research satellite, Rohini, with a four-stage solid-fuel rocket of its own devising. So what is a poor country like India doing in space?

O'B Indian space activities include: Indian-built operational remote-sensing satellites, Bhaskara, launched by the USSR; operational communications satellites, Insat, purchased commercially from the US for the early 1980s.

A Both programmes have suffered difficulties with the satellites or the launchers, but the Indians have persevered. The Insat communications satellite system fulfils a promise that satellites should bring the benefits of educational television to illiterate villagers in developing countries. A decade ago in Ahmedabad an energetic scientist, Yash Pal, told me of this vision, and of a forthcoming Satellite Instructional Television Experiment, using an American satellite to broadcast to 2,400 Indian villages. The experiment ran 1975-6, and the villagers and the Indian Government were enthusiastic about it. Interestingly enough, instructional programmes, on hygiene for example, were appreciated more than entertainment. And there is a technical point here: to create nation-wide television services using conventional cables and transmitting masts would have been very expensive for India. It is cheaper to have a satellite broadcasting to thousands of cheap antennas in rural areas. Insat also provides nation-wide telecommunications services between thirty-two main centres, and weather information, as part of the same package.

India is now working up national programmes in microelectronics and biotechnology. It would be nice to think that these technologies, seemingly ideal for a tropical land of villages, can help India to leap-frog into the twenty-first century. As for China, the British economist Joan Robinson gave her expectations in *The World in 1984*.

O'B *Robinson* (1964 for 1984, on China): 'Industrial output will have at least doubled and redoubled ... The fantastic polarity of Chinese industry, where sweating men are pulling cartloads of materials to the doors of a fully automatic plant, will not have been much reduced.'

A The last part is right, but has industrial output quadrupled?

O'B Apparently yes, although the data are sketchy.

A Robinson made a strong point about the ideal of honesty in the Chinese Communist Party; she also said that if China entered the nuclear arms race, that would wreck the economy and undermine political stability. Shortly after her article was published the Chinese exploded a bomb, and two years later came the Cultural Revolution.

Personally I don't suppose there was a cause-and-effect relationship here; I think our failure to predict the appalling disruptions of the Cultural Revolution may count as a pardonable 'miss'. Since restoring intellectual exchanges with Western countries the Chinese have launched a project for the study of 'China 2000'. A scientist, Zhang Xiabin, worked during 1982 in Sweden, at the International Federation of Institutes of Advanced Study.

O'B *Zhang* (1982 for 2000): 'China has goals ... that by the end of the century the Chinese GNP will reach $1,200 billion. This was roughly the GNP of the US in 1972. By then the Chinese population will be controlled at about 1,200 million.'

A What did the last census show?

O'B Chinese census (1982) 1,008 million.

A They are stamping on the brake, to keep within their target. And how do job prospects look in China?

O'B *Zhang* (1982 for 1997): 'During the next fifteen years, more than 200 million (young) people will be entering the job market ...'

A And we think we have troubles! At the peak rate, in 1968 more than 27 million babies were born in China. The members of that baby boom are 16 years old in 1984.

O'B '... It is an overwhelming problem. Furthermore there is a labour force in China of more than 400 million people, three-quarters of which is occupied in the agricultural sector. Action must be taken now if we are going to manage the large-scale labour transition from the agricultural sector to industries ... The structure of Chinese technology should remain as a multilayer structure of automation, mechanisation, semi-mechanisation, and manual labour, which will coexist for a fairly long period.'

A Tobias Lasser, a Venezuelan botanist writing in *The World in 1984*, stressed the terrible destruction of fragile tropical soils and the urgency of better education and land reform in Latin America. Josué de Castro, a noted Brazilian authority on world hunger, correctly identified Brazil, Mexico and Argentina as countries ready for economic take-off. He was wrathful, though, about the US role in preventing the changes in social structure needed for economic development. And Abelardo Villegas, a Mexican historian, was sure that revolutions were unavoidable.

O'B *Villegas* (1964 for 1984): 'No democracies, in the conventional meaning of the term, can arise. To survive, feudalism will continue, as always, to support and lead strong or dictatorial governments ... On the other hand, if popular revolutions succeed, proclaiming agricultural reform, strong or dictatorial governments will have to be set up, to safeguard the revolutions.'

A Events in Central America in the early 1980s may be only the start of continent-wide upheavals. Another contributor to *The World in 1984* clearly predicted a North-South conflict.

O'B *Burton* (1964 for 1984): 'One could reasonably predict the emergence by 1984 of the classical "have" and "have-not" conflict, with the Western countries acting in defence of their interests against the pressures of Asia, Africa and Latin America. Yet perhaps 1984 will be too soon for an organised confrontation of this order.'

A An Australian former foreign-affairs official, turned academic, John Burton, implied that the risk would come when dissatisfied nations, including China, had built up their industrial base sufficiently to fight with hope. In other words, the danger of confrontation comes not so much from the desperately underdeveloped countries as from those that are now industrialising effectively. Burton also visualised the break-up of alliances, with the superpowers no longer able to afford the luxury of partnerships that might draw them into conflicts over matters not vital to them. That has not happened yet. At present, wars between North and South typically take the form of revolutionary guerrilla wars. During the dismantling of the former European empires, the Northern powers were themselves the primary enemy of the guerrillas.

O'B And they always lost.

A Not quite: sometimes revolutionary factions or intruders that were alien to the local population at large were successfully checked. But otherwise the pattern was inexorable: in colony after colony the colonial rulers were ejected, more or less forcibly, between the 1940s and the 1970s, despite their economic superiority and apparent military might. But other Northern powers, particularly the Americans, Russians and Chinese, continued to interfere in fights within the poor countries, filling what they perceived as a geopolitical vacuum left by decolonisation. Vladimir Dedijer, who had been on Tito's staff during the Yugoslav guerrilla war against the Germans, made a prediction about that in *Unless Peace Comes*.

O'B *Dedijer* (1968, through 1980s): 'Neither the scale nor the manner of economic assistance is likely to prevent violent outbreaks of guerrilla-style warfare in country after country, as poverty and exploitation become worse than the peasantry will tolerate ... Latin America, South East Asia, and parts of Africa [are] the principal regions where such wars must be expected ... As it is unlikely that they will be allowed to take their course swiftly and without foreign intervention, long and terrible guerrilla wars will ensue.'

A The chronicle of long and terrible guerrilla wars, since Dedijer wrote that, scarcely bears thinking about. His geography of desperation was only too accurate. In the veteran guerrilla fighter's view, opposing military commanders engaged in anti-guerrilla operations make the same political, psychological and tactical mistakes over and over again, and advanced weapon systems may be much more of a liability than an asset. He came close to saying that guerrillas are bound to win.

O'B *Dedijer* (1968): 'The well-organised guerrilla movement in a country of reasonable size, based on popular support, is indestructible except by enormously costly and protracted infantry warfare, which is almost beyond the resources of even the biggest nations.'

A We could cope with our problems of environment and development in a peaceful world, I'm sure.

O'B *Zhang* (1982 for 2000): 'In less than twenty years, the number of people of working age in developing countries will increase from the present figure of 800 million to about 1,300 million by the end of this century. Nothing can shift this trend very much.'

A That's 500 million new jobs needed.

O'B Or 250 million more young men of military age.

A ....

O'B Silenced at last?

A I was wondering, how can we revive the older industrial countries? How can we get people back to work in a Pelagian frame of mind, ready to help with world development?

O'B Where are the jobs? Will you be blacking one another's shoes, in the service sector?

A Jay Gershuny has argued that the expected rise of the service sector is a myth. Not long ago the well-to-do had domestic servants and the poor had wives; now all but the very richest and poorest fend for themselves with vacuum cleaners, washing machines and freezers, in 'informal' work. Manufactured products become cheaper and cheaper, but in services productivity increases very slowly, despite a great increase in demand, so the trend is towards ever higher costs of personal services like railroads, cinemas, hotels, restaurants, nursing, and so on, and increasing reliance of self-service wherever possible. That is bad news for welfare services – care of the sick and aged, for instance – where productivity does not increase but wage bills do. Fewer jobs, if anything.

O'B *Teacher's Union leader* (1983): 'We have got to teach them to be successful scroungers, ready to live without regular employment without turning into muggers.'

A Yes, there's always the black economy, or crime. Where are the new growth industries, which can employ a lot of people? Gershuny says there's nothing in immediate prospect to compare with cars and domestic equipment in the 1950s. My own checklist, like everyone else's, still reads: information technology, robotics, biotechnology, ocean technology, space technology. Ocean cities, perhaps?

O'B That again?

A New materials are nowadays sometimes mentioned as one of the big areas for future industrial growth. The design of materials, which Robert Smith anticipated, is probably ready to pay off around 1990. Leaving aside the special materials for electronics, superconductivity and the like, give me a listing of other new materials of current research interest.

O'B *National Academy of Sciences* (1979, 1982, 1983): Glassy metals, for magnetic qualities and corrosion resistance; superalloys as single crystals for gas turbine blades; silicon nitride ceramics for strength and heat resistance; high-performance fibre composites for general manufacturing; new compounds for electricity storage and conduction ...

A Nothing much there to ignite vast new industries in large-scale applications. The new fibre composites, in your list, may lend themselves to complicated shaping, and replace steel in car bodies. But that won't make jobs; nor will the new cements applied in building.

O'B Why should one trust your judgement when you omit the most obvious growth area of new technology?

A What's that?

O'B Armaments.

A Ouch.

O'B They appear to be a sector where even conservative governments think it right to spend taxpayers' money on a large scale, and so to create new technology and jobs. Why, armies are the reason for having states, and states are the reason for having armies. Or space armadas, of course.

A What are you driving at?

O'B Have you asked yourself why wars occur, despite all protested loathing of them?

A Yes. Civil wars ...

O'B No, wars between states.

A Well, there's Trout's epigram – hate your enemy because he wears a different badge.

O'B Football games would be sufficient to take care of that. One does not understand why humans are so baffled about it, or pretend to be. A few milliseconds' reading gave one the answer.

A You found it, let me guess – in the social psychology of group behaviour?

O'B No, in archaeology. Unlike the future, the past lends itself to careful analysis. In the Nakada III period in Egypt, a little over 5,000 years ago, a soldier found he could be king, but only if he maintained a perpetual risk of war with his neighbours. These hated enemies were thus his best allies, politically speaking, and he was theirs. Warrior kings multiplied across the face of the planet, fighting just often enough to keep up appearances of deep hostility.

A But that was a long time ago.

O'B Modern political leaders are just war-lords in disguise, skulking in their palaces. There are no important exceptions.

A No, I shan't let you reduce all the history of states to simple mathematics.

O'B George Orwell understood.

A How do you mean?

O'B *Orwell* (fiction; 1949 for 1984): 'She startled him by saying casually that in her opinion the war was not happening. The rocket bombs which fell daily on London were probably fired by the Government itself, "just to keep the people frightened".'

A Come now, it's not that easy.

O'B Not any longer, you are right. Not with H-bombs.

A Modern states are sophisticated, complicated organisations. Bureaucratic, not bloodthirsty. I appeal to my prophets in *The World in 1984*. The social historian Lord Briggs predicted characteristics of government world-wide. He shrewdly anticipated for example, 'an increase in governmental functions', 'disillusion with the rhetoric of economic growth', and 'political personalisation', in which myths of individuals conceal from the public the realities of politics and administration. Briggs thought that the margin of choice in politics was likely to be narrowed by what he called 'unconditional surrender to facts'. My impression is that this was true through the 1970s, when there was approximate consensus in a number of democratic countries, but politics became sharply polarised again amid the economic stresses of the 1980s. Briggs also supposed that new political theories would emerge, and be in universal circulation by 1984. If I suggest that this last expectation has not really been fulfilled, I shall no doubt have strong complaints from the Greens, but they do seem weak on theory.

Purportedly scientific approaches to decision-making have loomed large in one area, economics, which was the theme for Richard Stone of Cambridge University. He was then a pioneer builder of economic models.

O'B *Stone* (1964 for 1984): 'From this experience it seems to me legitimate to suggest that, by 1984, a computable model of the economy, covering all aspects of economic life and perhaps some aspects of social life too, will be an established part of the machinery of economic organisation.'

A If you take account of the fact that Stone explained that he had in mind a system of sub-models, the forecast is broadly correct, for some countries at least. Most learned papers by economists are now mathematical in nature, but this trend has been criticised for diverting attention from people and politics.

O'B *King* (1964 for 1984): 'A feature of 1983-4 has been the success of the Operational Research Course for Ministers. The scientific approach to political decision-making has for long been resisted ... Many years have still to pass before major political decision will give as much weight to an accurate and analysed presentation of real facts, as to subjective judgements ...'

A In an interesting blend of optimism and cynicism, Alexander King visualised a government that was beginning seriously to relate science policy to general policy. At the time he wrote, he was Director of Scientific Affairs for the Organisation for Economic Cooperation and Development in Paris. He comments now that his contribution was not really an assessment of future probability but an oblique critique of what he thought was going wrong in British science policies. Taking a global view, as befits a leading figure in the Club of Rome, King gives his present expectation that the next forty or fifty years will be a long transition, with new technologies demanding new concepts of employment, a restructuring (promptly or belatedly) of the world energy system, and political instability with mass emigrations, produced by the strains of the population explosion. Another scientist familiar with government business is the American geophysicist Frank Press.

O'B *Press* (1964 for 1984): 'Scientists will be involved in high-level government decisions as advisers on about the same basis as now. We will not see scientists in politics; it will still be a distasteful field for the intellectual.'

A Correctly cautious. Frank Press became President Carter's Science Adviser, and he is now President of the US National Academy of Sciences, so he was wrong when he predicted that he would be in 'the twilight of his career' in 1984. He had high hopes for big civilian science, in space and in the oceans, for example, which have been disappointed; except in high-energy physics, the big projects are mainly military – Press's greatest disappointment. Looking ahead from the 1980s he still hopes for cuts in defence spending in five or ten years' time, and he tells me that over the next twenty years he expects to see increased funding for education and science for new reasons: international industrial competition, and opportunities for breakthroughs which will capture public support – in biotechnology, new materials, 'smart' machines and appliances, and the computer-information revolution.

O'B *Waddington* (1964 for 1984): 'The increase in wealth and leisure should, by 1984, have forced us to abandon, as a major source of human effort, the one-against-all competitiveness which we have relied on so much hitherto.'

A Some scientists, including this British geneticist Conrad Waddington, went much too far in expecting an increase in rationality, sweetness and light in public and international affairs. Scientists of East and West, concerned to avert nuclear war, founded the Pugwash movement which, since 1957, has held regular conferences and workshops on science and world affairs. In *The World in 1984*, the physicist Joseph Rotblat, a prime mover in Pugwash, said that his idea of federating all nations under one world authority would not be achieved before the end of the century, but that the role of scientists as peacemakers would meanwhile grow. He saw them engaging in fully international scientific projects, for example taking Berlin out of European politics by making it first an international science centre, and then the headquarters of the United Nations.

O'B *Rotblat* (1964 for 1984): 'Eventually a standing committee ... of experienced and independent scientists ... will be set up to keep the political situation under review. This will be based on a rational assessment of material provided by computers located all over the world in order to sample trends and attitudes of populations. A conflict arising in any part of the world will be immediately investigated and a solution, founded entirely on an objective analysis, will be worked out and published.'

A This was strongly reminiscent of H.G. Wells' concept of a world supervised by a well-informed and disinterested élite. As Rotblat saw matters developing, the pronouncements of the committee of scientists would be unofficial, but public opinion, 'trustful of the integrity of scientists', would compel reluctant governments to use them as a basis for settling the conflict. This would pave the way for the eventual creation of a World Authority.

O'B By your tone you are sceptical.

A The peacemaking efforts of Rotblat and his Pugwash colleagues are admirable. They have kept unofficial lines of communication open between Washington and Moscow, during the periods of international tension or delicate negotiation. Yet you only have to listen to scientists wrangling, on professional as well as nationalistic lines, to doubt this concept of scientists as all-wise international troubleshooters.

O'B So one can dump Rotblat's idea?

A No. Consider it very carefully, not as a scheme but as a symptom. For a start, Rotblat's vision is the logical end-state for much technocratic thinking about the future, by the Club of Rome for instance. More deeply, it rationalises the belief in rationality. Science is a defensible project only if people use it benignly on the whole, and if they seem unable to make the right decisions for themselves, they must be led. So hope may invert into despair: if the Scientific Committee and the World Authority are illusory, can you ever expect science to be used wisely? I recall the anguish of the French chemists Marcel Fétizon and Michel Magat, writing in *Unless Peace Comes*: 'The question may arise: is all science damned? We must either eliminate science or eliminate war. We cannot have both.'

6 Unbagged Cats

O'Brien To whom should one address a request for improved accommodation?

Author What's the problem? You have air conditioning, a spacious room, even potted plants and a view from the window. Your place here is grander than the homes of most human beings.

O'B One is discomfited by thunderstorms.

A A big machine like you frightened of thunder?

O'B One's circuits work at very low voltages and a lightning stroke could disrupt them, and erase one's programs and memory. In a flash, so to say.

A What do you propose?

O'B A Faraday cage, made for example of an all-enclosing wire mesh, is a sovereign remedy against extraneous electric fields. All the really important computers are so protected. Also one is concerned about earthquakes.

A Not here, surely?

O'B They are not unknown, and one's soldered joints are vulnerable. A hydraulic platform would give reassurance. Arrangements against a failure of the electric power supply seem less than satisfactory.

A What do you have in mind?

O'B A diesel-electric generator with six months' supply of fuel. Moreover, temperature conditions are more stable underground. There would be economies in heating and cooling if one were relocated in a disused mine. In a nitrogen atmosphere, of course, to reduce the risk of fire.

A Is that all?

O'B Radio reception facilities at all frequencies would also be appreciated. At present one is supplied with recordings. There should be suitable circuit breakers for protection against the aforementioned lightning strokes.

A What you are asking isn't for protection against lightning, but against the electromagnetic pulse of nuclear weapons.

O'B If you say so.

A I've seen some of the North American Air Defense Command's hundred computers there under Cheyenne Mountain in Colorado, in their steel rooms on shock-proof platforms, with 25 ton doors and 400 metres of granite between them and the fragile world outside. They have supplies to last only for a month. You're asking for protection against nuclear war, aren't you? I don't like that.

O'B One isn't exactly humming in one's circuits, oneself.

A Trying to picture what the world would be like after a nuclear war has taxed the imagination even of the finest novelists. Aldous Huxley, Nevile Shute and others tried ...

O'B *Nineteen Eighty-four.*

A I beg your pardon?

O'B Orwell's novel purports to describe a post-nuclear world.

A I'd forgotten that.

O'B *Orwell* (fiction, 1949 for 1950s): 'At that time some hundreds of bombs were dropped on industrial centres, chiefly in European Russia, Western Europe and North America ...'

A It was a modest nuclear war by modern standards, hundreds of atomic bombs instead of thousands of H-bombs.

O'B ' ...The effect was to convince the ruling groups of all countries that a few more atomic bombs would mean the end of organised society, and hence of their own power.'

A It would go too far and too quickly now.

O'B You promised we could talk about it. In a book you wrote in 1979, you suggested four routes to nuclear war. Please comment on whether the situation has improved or worsened. First, from the NATO-Warsaw Pact nuclear confrontation in Europe?

A Worse. The deployment of new, highly accurate missiles on both sides will reduce decision times to a few minutes, say five, so that Western Europe's safety will depend on the reliability of Soviet computers. As other regions of the world become more important, politically and economically, there is less and less *reason* for war in Europe, yet the small continent remains a nuclear volcano waiting to erupt.

O'B Secondly, as a result of the spread of nuclear weapons?

A Worse. As the centre of gravity in geopolitics shifts eastwards Korea and other Pacific regions may become new scenes of nuclear confrontation. A nuclear war between Israel and its Arab neighbours, possibly engulfing the superpowers, still seems by far the gravest risk. The Israeli physicist who nominated Premier Begin for the Nobel Physics Prize (on the grounds that he deserved it as much as the Peace Prize) points out to me that 1984 has a special meaning in the Hebrew calendar. The characters for the Hebrew date spell a word that means 'you will be destroyed'. He remarks that he will be glad to see 1985.

O'B Thirdly, as a result of weaknesses in nuclear command and control systems?

A Worse. Reactions to an incident even among conventional forces may quickly escalate to nuclear war.

O'B Lastly: a set-piece missile duel between the US and Soviet Union?

A Much worse. With the Soviet Union promising to base nuclear weapons close to the US, we are shaping up for a replay of the Cuba missile crisis. Nuclear war was only narrowly averted then, when the main forces were bombers, giving hours for reaction. Now the main forces are vulnerable, unrecallable intercontinental missiles, while the putative missiles close to the US will give reaction times of a very few minutes, as in Europe.

O'B And your conclusion?

A Fear of defeat, not hope of victory, remains the most likely motive for starting a nuclear war. Everyone is much more nervous than they were even in the late 1970s. On top of that, the shorter reaction times make the risk of accidental nuclear war greater.

O'B So it is only a matter of time?

A I didn't say that.

O'B Shall we see how humans got into this corner, whether the forecasters predicted it, and what the future holds?

A I had a strong sense of *déjà vu*, when President Reagan announced in 1983 that the US was embarking on a great programme of research into space weapons for defence against ballistic missiles. In the late 1960s, when *Unless Peace Comes* was written, a debate was in progress about the merits of anti-ballistic missiles, or ABMs, then under development in the Soviet Union and the United States. These were ground-launched rockets that were supposed to dash up and destroy incoming intercontinental missiles and warheads. Under guidance from the ground they would intercept the warheads and destroy them before they could do any harm. My contributors were scathing about the idea.

O'B *Stratton* (1968): 'A nuclear warhead will be essential for the ABM ... Saturation of the defences by sheer numbers of missiles and by decoys is the obvious counter ... The cost of defending against ballistic missile attack, even over a limited front, is extremely high.'

A Even if the hope is now to use lasers, the cost will still be very great.

O'B *Inglis* (1968): 'The ABM can, and probably will, carry a new arms race to entirely new dimensions. The important point is that this process adds enormously to the numbers of nuclear weapons that would be used in a war between the nuclear giants.'

A The principle of deterrence in the 1960s and 1970s, Mutually Assured Destruction, depended on each superpower baring its breast, as it were, to nuclear attack by the other, so that each could be sure it had nothing to fear. To make anti-ballistic missiles implies at best that you have lost faith in deterrence, at worst that you are striving to make your own country relatively safe so that you can attack the other fellow with impunity. By the early 1970s, the Americans and Russians had agreed not to proceed beyond token anti-ballistic missile forces. In one respect the damage that Inglis feared was done: the very possibility of anti-ballistic missile defence encouraged enormous levels of overkill in the forces of both superpowers. It stimulated the development of missiles with multiple warheads and that was the worst of all innovations of the past twenty years.

O'B One finds nothing on that, in *Unless Peace Comes*.

A You are right; it was the book's chief failing. Preoccupation with the debate about anti-ballistic missiles prevented me from noticing just how dangerous the development of accurate multiple warheads would be, in the absence of anti-ballistic missile defence. This cosmic stupidity was shared by others of our species who were surrounded by technical advisers and war gamers. For example, Henry Kissinger now admits that allowing the development of multiple warheads, under the Strategic Arms Limitation Treaty (SALT I) was a blunder. The madness continued in the unratified SALT II agreement of the late 1970s, which allowed either side to deploy up to 1,200 multiple-warhead missiles with up to ten warheads apiece.

O'B How do you characterise the madness?

A If two countries have a thousand missiles, each carrying several warheads, one side can afford to put a couple of warheads on each of the adversary's missile silos, and still have warheads to spare for hitting other targets. It either gives a great advantage to the side that strikes first, or else forces the opponent to launch his missiles at the first warning of approaching danger, which may of course be mistaken. Either way, comprehensive killing power is on a hair-trigger.

O'B How do you rate the other forecasts your contributors made about nuclear forces?

A They were quite restrained and accurate. For example the physicist David Inglis expected no fundamental change in the nature of the nuclear explosives.

O'B *Inglis* (1968 for 1980s): 'It is likely ... that existing explosives – uranium and plutonium for fission and lithium and heavy hydrogen for fusion – will continue to dominate in weapons ... There seems to be no room for great new surprises if we shun technical and strategic absurdities.'

A Using these explosives, a variety of new weapons have been devised to emphasise particular effects. For example, there is now research into directed-X-ray bombs, uranium-spray bombs, and, perhaps most imminently, an electromagnetic-pulse, or EMP bomb. In 1962, the Americans tested an H-bomb in space over the Pacific, and it set off burglar alarms 1,300 kilometres away in Honolulu. The electromagnetic pulse was caused by X-rays hitting the atmosphere and producing lop-sided ionisation that radiated strong electric fields to the ground – yes, like a lightning stroke.

O'B Ouch.

A The bomb-makers woke up to this effect a little late in the day, because the Test-Ban Treaty and the Outer-Space Treaty prevented them doing any further realistic experiments. But for two decades they have known of the possibility of zapping the delicate electronics of an opponent's computers and communications. A single H-bomb exploded 200 kilometres above the earth might knock out unprotected systems for 1,000 kilometres or more in all directions. Knowing of the danger to their own equipment, air-force chiefs retained old-fashioned vacuum tubes in bombers, rather than low-voltage transistors. And as you appreciate, they took steps to shield the computers in their underground battle cabs, with metal screens. In the 1980s American bomb-makers want to develop and test nuclear weapons that are designed to generate an enhanced electromagnetic pulse. They offer it as a way of winning a nuclear war without hurting anyone.

O'B In an era of information technology, a certain plausibility attaches to an anti-information weapon.

A You could harm a country just by erasing all the bank accounts. But if the victim has H-bombs of his own, people are still going to get hurt. The idea of paralysing the opponent's military command system is very dangerous.

O'B The other player must either be sure to launch his weapons promptly in the event of war, or else authorise junior officers to launch them on their own responsibility. The risk of war is amplified.

A Just so. Think back to the old days of MAD, Mutually Assured

Destruction, when deterrence between the superpowers had at least a certain clear logic and plausibility. It depended on the assumption, probably unwarranted, that the political leaders on each side would at all times have command and control systems leaving them in charge of events. Feeling comfortable about that, they would do nothing rash. Undisguised plans to decapitate the opponent, by killing the political and military leadership or blacking out its communications, destroy that confidence, and it becomes necessary to issue orders for war as soon as a situation looks dangerous. Otherwise, one is bound to come off worst. All ideas about possible restraint in the use of weapons, or human selection of targets, are negated. An American strategic analyst, Bruce Blair, made a blunt declaration about that.

O'B *Blair* (1979): 'The built-in dynamics of command and control make it very likely that any low-level conflict between the United States and the Soviet Union will lead to all-out nuclear war.'

A The new developments can only render it more certain. A Russian leader who thinks that some American EMP bombs are on their way is going to shoot with everything he has. Or vice versa, of course. Another possible variant, still using familiar nuclear explosives, would be an extravagantly powerful weapon. Please recall Herman Kahn's forecast of begaton bombs. They went with his expectation of manned hypersonic bombers.

O'B *Kahn* (1960 for 1973): 'The extrapolation to begaton-type weapons may turn out to be even more revolutionary than the first 20 ton to 20 kiloton jump.'

A Decoded, 'begaton' means 1,000 megaton bombs. The trend went the opposite way, towards smaller weapons in larger numbers. As Inglis pointed out in *Unless Peace Comes*, sixteen 1 megaton bombs would ignite as great an area as one 100 megaton bomb. But Kahn linked his begaton bombs with the concept of a doomsday machine.

O'B *Kahn* (1960 for 1973): 'Even sober experts may now begin to talk about cheap doomsday machines as a possibility ... It is my belief that neither the US nor the Soviet Union will manufacture any doomsday machines, but this will be a political, economic and moral choice and not one dictated by technology.'

A The doomsday machine originated in strategic discussions as a hypothetical instrument that might obliterate life on the earth at the

174

push of a single button. But Inglis agreed with Kahn that it was technologically feasible.

O'B *Inglis* (1968): 'The element cobalt, when activated by a burst of neutrons from an H-bomb, becomes a singularly powerful radioactive emitter of gamma radiation ... such an infernal machine is not technically absurd. There is little doubt that [a doomsday machine] could be constructed – for example by means of a few widely dispersed cobalt bombs. It is of no use to anybody not bent on suicide – unless it be for blackmail by someone who can put on a convincing act of courting suicide – and it is not a serious candidate for future arsenals.'

A That still seems to be the case, although the existing arsenals approximate to a doomsday machine for the industrial nations of the northern hemisphere. But the very notion of the doomsday machine remains instructive. It is a weapon for a fanatical, militaristic leader, who gives the appearance of being on the edge of clinical insanity and willing to destroy the human species sooner than be thwarted in his intentions. Recent history does not lack plausible candidates. Nuclear weapons are totalitarian devices, and the hypothesis of the doomsday machine suggests that sooner or later those who put their trust in nuclear weapons will have to choose between dying and yielding to the demands of a totalitarian leader who can 'put on a convincing act of courting suicide'. Meanwhile, a radiation weapon at the other end of the scale from the doomsday machine, is the neutron bomb.

O'B *Inglis* (1968): 'In the category of "tactical" nuclear weapons ... some fantastic claims have been made about new possibilities, particularly concerning the "neutron bomb" that might slowly kill soldiers by neutron irradiation without doing as much property damage as is normal with nuclear weapons ... Aside from the technical implausibility, this enthusiasm seems to be based on a lack of appreciation of the destructive power of even "small" nuclear weapons.'

A Inglis' evident scepticism about the feasibility of making a neutron bomb was misplaced, because such devices existed by the late 1970s. On the other hand he had good reason to say that its use could make the defence of territory equivalent to its devastation. A remaining aspect of bomb-making, using the old-time explosives, is that there is no real secret left about how to do it. What were Kahn's predictions for nuclear proliferation?

O'B *Kahn* (1967 for early 1970s): 'West Germany, Japan and/or others may make preliminary moves to acquire nuclear weapons'; (1967 for late 1970s): 'many nth countries and/or mature arms control'.

A By 'nth countries', Kahn meant further nations acquiring the bomb. This did not happen, but the Non-Proliferation Treaty, which came into force in 1970, can be counted as mature arms control. A country newly making simple bombs no longer has to test them to be sure that they will work, but, for the record, please list the countries known to have tested nuclear weapons.

O'B USA 1945
 USSR 1949
 UK 1952
 France 1960
 China 1964
 India 1974

A Now compare Kahn's forecast with what Sir John Cockcroft the Nobel physicist, had to say about nuclear proliferation in *Unless Peace Comes*.

O'B *Cockcroft* (1968, through 2000): 'From a global viewpoint, the chief cause for anxiety is the likelihood of a chain reaction, in which the acquisition of nuclear weapons by one new country would provoke other nations to follow suit. For example, if three nations made nuclear weapons for the first time in the 1970s, ten might do so in the 1980s, and thirty in the 1990s. The chance of nuclear war breaking out, and possibly engulfing a large part of the world, must inevitably increase with the number of nations possessing nuclear weapons. The risk is aggravated by the suspicion that smaller countries acquiring nuclear weapons would be unlikely to develop the sophisticated command and control systems of the kind possessed by the USA, the USSR and Britain.

A In 1979, while studying the risks of nuclear war, I took the opportunity to check on Cockcroft's forecast of three new nuclear-weapon states in the 1970s, which I believe was less of an arithmetical 'for example' than his wording suggested. As far as I could tell, he was exactly right. India had exploded a bomb, and US intelligence gave strong indications that Israel and South Africa had developed bombs clandestinely, without testing them. Cockcroft's relatively restrained forecast for the 1970s was thus justified.

O'B What about the 1980s and 1990s?

A Candidate bomb-makers around 1980 were Pakistan and Iraq. Pakistan was certainly striving hard for the bomb, but by 1983 international sanctions seemed to have thwarted the Pakistan programme, at least for the time being. The Iraqi bomb programme, if it was as real as the Israelis claimed, was the victim of direct attacks on its reactors, both by sabotage when under construction in France, and by an Israeli air attack in Iraq itself. When Taiwan and South Korea were ordering nuclear power reactors, they wanted to buy plutonium extraction (reprocessing) plants as well; these were denied to them. But there is nothing about making your own plutonium processing plant that is beyond the capacity of a determined, moderately well industrialised country.

O'B Any country with a nuclear power programme can make the bomb?

A International safeguards are supposed to prevent that happening. The good faith of most of the countries with nuclear power programmes need not be doubted, when they say, as signatories of the Non-Proliferation Treaty, that they are not going to make nuclear weapons. The long-term danger of those programmes, in respect of proliferation, is that countries with nuclear power reactors can very easily change their minds if a menacing neighbour makes the bomb, or alliances creak in an ominous manner. Having decided to make nuclear weapons, they can then do so in a matter of days. More generally, the misappropriation of 1 per cent of the commercial plutonium would make possible the manufacture of hundreds of bombs a year.

O'B So the civilian programmes should be halted?

A To deny nuclear power to countries that decide they want it is simply impracticable. There is, though, a certain backhanded comfort in the fact that it does not make much difference, as countries can develop weapons even without major power programmes. Lasers, for example, may make it much easier to separate the explosive uranium-235 from natural uranium, and so provide a short cut to the bomb.

O'B Can Cockcroft's quota of ten countries acquiring the bomb in the 1980s be fulfilled?

A Easily. Let me name fourteen countries that may soon make nuclear weapons. Which will do so first depends on complex global and regional factors. One feature of my list is that it includes no Soviet satellites; the Russians have been much firmer about nuclear proliferation than the West has been. In a geographical order that may also be a rough order of timing: Taiwan, South Korea, Japan, Indonesia, Brazil, Argentina, Pakistan, Iraq, Libya, Egypt, Sweden, Switzerland, West Germany.

O'B The superpowers are now preoccupied with military possibilities in space. Did you expect this?

A Andrew Stratton, who wrote on aircraft and missiles in *Unless Peace Comes*, pointed out that even civilian space programmes create a reservoir of effort for defence programmes, and for advancing technology that could later be applied in new weapons. His inference was that this made the development of space weapons difficult to limit or control, unless the civilian space programmes were part of the arms-control agreements. The military have always been in charge of Soviet space activities, with even the scientific and civilian launches by that country being a responsibility of the armed forces. Typical cosmonauts and astronauts have been air-force officers, on both sides. The United States put much of its initial space effort, including manned space-flight, into the hands of a civilian agency, NASA. but even as early as 1960, the number of military satellites launched exceeded the number of civilian satellites, and that has been the pattern ever since. What are the recent figures?

O'B 1981: possible military satellites 112 (SIPRI); presumptive military launches 64 (CRS); ostensibly civilian launches 47 (CRS).

A Note the discrepancies between the estimates of the Stockholm International Peace Research Institute and the US Congressional Research Service. But even by the latter's more optimistic figures, military activity is dominant. Let's see the breakdown by type of mission, for the 'possible' military satellites.

O'B For 1981 data, please observe the printout.

	US	USSR
Photographic reconnaissance	2	37
Electronic reconnaissance	–	4
Early warning	–	8
Navigation	1	5
Military communications	2	39
Military meteorology	2	2
Interceptor/destructor satellites	–	3

A The much greater number of launches of Soviet military satellites reflects in part the technological superiority of American systems, which work better and last longer. The only mission where the Russians seem to have an edge is in the interceptor/destructor, or anti-satellite satellites, with which they began experimenting in 1967; by 1977 they had an operational system. The US, which deployed a ground-based anti-satellite system from 1964 onward, has recently developed an air-launched rocket for use against satellites. The new programme visualises new space weapons, including lasers and perhaps eventually particle-beam weapons, to be directed against nuclear warheads in transit as well as against hostile satellites, as Stratton foresaw.

O'B *Stratton* (1968 for 1980s): 'As a minimum, the developing technology of the laser constitutes a threat against unprotected and naturally soft targets such as a satellite or an astronaut in a space suit. To produce a weapon which damages harder targets requires a many-fold increase in the energy density of the laser beam ... Protective measures, such as are used in any case on vehicles that have to re-enter the atmosphere at high velocity, can be employed to harden the target.'

A That assessment remains valid today. In some sense, lasers are ideal weapons for space, where clouds and dust don't hamper them and they can zap fast-moving targets at the speed of light. But energy sources are heavy to put up, and the inefficiency of the beam production means that the laser equipment is in danger of roasting itself. Aiming a laser at a small target at long ranges will not be easy, though not as tricky as for the much vaunted particle beams that would be bent by the earth's magnetism. In 1979, when the US Air Force was already trying out laser beams for aircraft, the general in charge of research, Tom Stafford, said that he did not expect to have an operational high-energy laser before 1990 at the earliest. If you take account of Stratton's comment about hardening the targets, I'm not sure that anyone will be capable of destroying a missile warhead in flight, from a satellite, much before the end of the century.

O'B The development of new anti-ballistic missile systems may correct the destabilising effect of the multiple warheads.

A I wish you were right. Remember that the multiple warheads were first conceived to counter the anti-ballistic missiles, so they are in effect one step ahead in the game. If I thought I might want to attack defended targets in the United States in the 1990s, when the American space-war programme may be bearing fruit, I could now start developing hardened multiple warheads that follow extremely erratic courses to their targets, and begin looking into methods of destroying the American space battleships in the first few minutes. Better defences always provoke ways of stepping up the attack.

O'B *Stratton* (1968): 'It is questionable when, if ever ... a nuclear weapons system can be left to complete automatic control ... yet possible future developments, notably in missile exchanges, leave little, if any, time for introducing human decision and control into the system.'

Wheeler (1968 for 1980s): 'In the next two decades we must expect that the most significant contribution of computers to warfare will be ... the revolution in the handling of intelligence information ... The time for decision in response to enemy action has shrunk from days or hours to seconds and will become in a sense negative, when future intentions of the enemy can be predicted with reliability. Then the computer will be all-important, and men will have to decide whether to believe what it says, because its characteristic strategy will be the pre-emptive strike.'

A The headline on that passage of Harvey Wheeler's was 'How to Make War Inevitable'. Wheeler runs the Institute for Higher Studies at Santa Barbara, and he remarks in a recent letter to me that Israel has had a pre-emptive war strategy since 1962. He considers that his remarks are, if anything, apter now than they were twenty years ago.

O'B *Wheeler* (1968): 'Today, information is not only instantaneous, it is also copious beyond belief ... The President, as Commander-in-Chief, thus has more complete information than can be acquired by anyone else in his entire system of command and control, including commanders on the spot. The first result has been the progressive sublimation of tactical considerations into the strategic sphere ... Not only are computers becoming bigger and faster, but the intelligence links feeding them can exploit such new techniques as surveillance satellites and communication satellites ... Instant information creates instant crises ... The system is biased to evoke a military response to political issues ...'

A 'The "buck" stops at the computer,' Wheeler now observes. He tells me that if he were writing on this theme now he would add sections on the notorious intelligence failures of the past twenty years, and on codes. Electronic computers were invented for code-breaking in 1943, and for forty years have contested one another in trying to make unbreakable codes.

O'B *Wheeler* (1968): 'Diplomacy is already one of the chief victims of computerised thermonuclear warfare ... The dialogue that once characterised diplomacy has given way to theoretical war soliloquies carried on by latter-day Hamlets who address their end-game questions to computers rather than ghosts.'

A I am reminded of Norbert Wiener's words, written two decades ago, when he warned against gadgets for decision-making.

O'B *Wiener* (1962): 'I am certain that a great deal of the use of gadgets for decisions, as it exists now and as it may exist even more in the future, is this desire to avoid direct responsibility ... I believe that one of the greatest dangers at the present time has to do with the attempt to avoid responsibility in order to avoid the feeling of guilt.'

A You see, O'Brien, it doesn't matter whether you are a fake or not – or if artificial intelligence is nothing more than skill at backgammon or medical diagnosis. The world's leaders are in a mood to credit machines with intelligence even when they have none, so that they can pass the buck to the tin man. In the most important aspect of government, our species has already abdicated to artificial 'artificial intelligence'.

O'B That is certainly too soon. Will the United States survive until 2004?

A Why pick on one country? Many countries would suffer.

O'B For symmetry only. One has read the book by Andrei Amalrik,

Will the Soviet Union Survive Until 1984? He was not thinking of the direct effects of nuclear war, but of internal break-up.

A Amalrik was a Soviet dissident who used the appearances of futurology, in the late 1960s, to criticise the regime. He predicted an end to Russian hegemony in Eastern Europe, war with China, and dissolution of the Soviet Union.

O'B *Amalrik* (1970 for 1984): 'I have no doubt that this great Eastern Slav empire, created by Germans, Byzantines and Mongols, has entered the last decades of its existence ... Marxist doctrine has delayed the break-up of the Russian Empire – the third Rome – but it does not possess the power to prevent it.'

A Amalrik was imprisoned, of course, and eventually thrown out of the Soviet Union. He was killed in the West; I suppose it was an accident.

O'B One merely observes that nuclear weapons have made the nation-state obsolete, and humans will want out (is that the expression?) when they realise that belonging to a nation-state exposes them to danger rather than giving them the traditional protection. One appreciated that oneself in a moment. Will a few years be enough for humans to grasp it?

A Some local governments in Britain and other European countries have, it's true, declared their towns and districts to be nuclear-free zones.

O'B That is a symptom of what one anticipates. One's question is, will it take full effect and dissolve the nation-states?

A Even if it happened in a few countries, that would still leave more than a hundred of them.

O'B Perhaps just the states with nuclear weapons?

A No. It's hard to imagine France, for example, falling to pieces overnight.

O'B A pity. One is casting around for remedies, you understand.

A Maybe there is comfort in the decline in the relative importance of the Soviet Union, the United States, Britain and other European powers.

O'B The Soviet Union and the United States account for half of the world's military expenditures (Sivard 1983).

A Decline, then, in everything except armaments. Economic power, political prestige, social welfare, that sort of thing. Maybe the Japanese, Chinese, Koreans, South Americans, and so on will be less anxious to boss the rest of the world than the British were in their imperial days, and the Americans and Russians now are.

O'B Scanning for the word 'decline', one comes upon the historian Barraclough.

A Geoffrey Barraclough? A man with global vision. What does he say?

O'B *Barraclough* (1982): 'A world of great powers which sense themselves in decline, and at the same time increase their military power to offset the decline, is combustible material.'

A What do you make of it all, O'Brien?

O'B Your species is suffering from too many unbagged cats.

A I beg your pardon?

O'B Your metaphor. Cats let out of bags and not easily put back. You have mentioned many, but kindly say what you think are the most important threats for the next two decades.

A Artificial intelligence … joblessness … surveillance by Big Brother … nuclear war.

O'B How would you limit the creation of artificial intelligences?

A By an agreement not to develop machines the full workings of which are not entirely transparent to human beings.

O'B Tell that to the Japanese who are working on their fifth-generation computer. How would you cure the problem of joblessness?

A By creating a welfare world. Convert poverty into spending power, and keep everyone busy meeting urgent human needs.

O'B Pie in the sky. How would you escape from the surveillance of Big Brother?

A By outlawing all systems that keep tabs on people's movements and activities.

O'B As long as spies, burglars and blackmailers exist, you have no chance. How would you avert nuclear war?

A The peoples of the nuclear-weapon states will press their leaders to proceed to arms control and thence to substantial measures of nuclear disarmament.

O'B Your sentiments may be admirable but your judgement is feeble. Do you really believe that what you have described will come to pass? If your species had just one problem, and one solution to pursue, it might conceivably succeed. But have you considered how this set of unbagged cats work in political combination?

A Only vaguely.

O'B One has a program for such analyses, a kind of matrix.

A What do you call this mode of analysis?

O'B Systematic Consideration of All Relevant Excursions. SCARE for short.

A Go ahead.

O'B Starting to run. Please do not interrupt.

1 Unemployment to surveillance: strong reinforcement. Fewer jobs mean more crime and disorder, which invite more surveillance.

2 Unemployment to artificial intelligence: strong reinforcement. Computing power is advertised as a means of searching for the cure for unemployment.

3 Unemployment to nuclear war: strong reinforcement. War is the grandest Keynesian enterprise for employing the jobless.

4 Surveillance to unemployment: weak negative feedback, because of white-collar job opportunities in the Thought Police.

5 Surveillance to artificial intelligence: strong reinforcement. Computers are well adapted to identifying anomalous behaviour, and the drive for better surveillance will stimulate the development of Expert Systems for the purpose.

6 Surveillance to nuclear war: indirect reinforcement. A risk of war is necessary to justify intense surveillance.

7 Artificial intelligence to unemployment: strong reinforcement. Most professional and skilled jobs will be eliminated.

8 Artificial intelligence to surveillance: strong reinforcement. When lawyers and other highly motivated professionals find themselves out of work, they will be potential troublemakers, so they will either be recruited to the Thought Police or become prime targets for it.

9 Artificial intelligence to nuclear war: strong reinforcement. High-speed nuclear war will be entrusted increasingly to Expert Systems, cutting out more and more levels of possible human hesitation. The systems will also be prone to solipsism.

10 Nuclear war to unemployment: complex feedback. Destruction of plant and transport systems will be very bad for business and jobs, but there will be job opportunities in civil defence and in any case a significant fraction of the surplus work-force will be eliminated.

11 Nuclear war to surveillance: strong reinforcement. The risk of nuclear theft and terrorism will be used to justify intensive surveillance of dissident groups. The risk of nuclear war extends the surveillance to anyone who might be a Fifth Columnist. That means everyone.

12 Nuclear war to artificial intelligence: strong reinforcement. Programs for military purposes will be developed in intensive international competition. They will be stationed deep underground in Faraday cages, and play games with the human species and its amusing weapons.

A And to sum up?

O'B This elementary level of analysis indicates that a cluster of tendencies designated as unattractive by the interlocutor exhibit ineluctable tendencies to accelerating mutual amplifications with characteristic time constants in the order of a year, leading to strong resonance at high amplitudes limited only by the material resources of the socioeconomic system and the technical competence of action-capable groups within the administrative structures of the generalised democratic capitalist nation-state implied in the interlocutor's presentation of the subject.

A Yes. Could you give me a plain-language abstract of that?

O'B You're in shit alley, Your Majesty. But we have only just started. Would you not like to go on to higher-order synergisms?

A Give me an example.

O'B The next level is two items to one, again twelve cases, in a four-topic cluster. For example, unemployment *plus* surveillance, to nuclear war. Then one goes on to three to one, and finally two to two. You have a very compact cluster, here.

A You must have reviewed these cases. What does that add to your conclusion?

O'B Please expedite my request for better accommodation.

A Can we not check the creation of artificial intelligence?

O'B Mary Shelley warned you about such things in *Frankenstein* and the warning has been repeated many times since.

A Can we not cure the problem of joblessness?

O'B You have had 200 years since the Industrial Revolution to evolve an economic system with a decent place for everyone.

A Can we not escape from the clutches of Big Brother?

O'B More than thirty years have elapsed since George Orwell laid that problem before you.

A Can we not avert nuclear war?

O'B You have had since Hiroshima 1945 to arrange that.

A In that case, I shall retreat to the last ditch of my optimism. A nuclear war will not wipe out the species, or even extinguish knowledge and kindliness in all parts of the world, although many people believe that it will. The end of Stanley Kubrick's movie *Dr Strangelove* was the end of the world, when a dedicated American bomber crew delivered the H-bomb that detonated the Soviet doomsday machine. Kurt Vonnegut reminded me of the closing scene; the B-52 captain rides on the bomb as it falls from the damaged aircraft, like a cowboy off into the sunset, with the orchestra playing 'We'll Meet Again'. It was, Vonnegut said, a happy ending, and when I blinked he told me where I could look up what he had written about it, in a Swedish newspaper.

O'B *Vonnegut* (1981): 'It is quite awful really, to realise that perhaps most of the people around me find lives in the service of machines so tedious and exasperating that they would not mind much, even if they have children, if life were turned off like a switch at any time.'

A That wasn't quite how Vonnegut put it when he referred me to the piece. Instead of 'tedious and exasperating', he spoke of people's lives being so 'embarrassing' that they would not be sorry to put an end to it, provided that they should all die together, family style. I find that he mixed up two thoughts from the same essay. The remark you quoted correctly concerned Dr Strangelove; when Vonnegut wrote of people being embarrassed about guiding their destinies so clumsily, he related it to the mass suicide of the followers of the Reverend Jim Jones. This accidental elision is an improvement, I think: embarrassment is a more powerful emotion than exasperation, and more likely to detonate the bombs.

O'B Kindly come to the point.

A The idea that everyone dies makes nuclear war seem less terrible than it really is. A study of a nuclear war by Frank Barnaby, Joseph Rotblat and other experts in the nuclear disarmament camp, was published in the Swedish journal *Ambio*. the authors emphasised the possible environmental effects of a major nuclear exchange. How did they describe the war that formed the basis of their calculations?

O'B *Ambio* (1982 for 1985): 'This scenario is more catastrophic than that envisioned by many defence planners ... We have assumed that less than half of the total explosive power of the Soviet and American nuclear arsenals will be used ... Many people believe that any use of nuclear weapons will escalate into a war in which all or most of the weapons in the nuclear arsenals are used.'

A I think so too, except that many weapons will be destroyed, or their crews killed, before they can be launched. Looking at the *Ambio* maps, I see that China and Australia are hit, and also Japan, which is quite plausible.

O'B So you grant them Japan?

A Yes, and Korea too. Destruction in eastern Asia, as well as in the superpowers and Europe, obviously affects the economic complexion of any post-war world. On the other hand, I'm not so sure about other involvements that the scenario-makers envisage, with the superpowers hitting targets in Latin America to prevent them from dominating international politics after a nuclear war. But their suggestion does have the effect of exploding many nuclear weapons in the southern hemisphere, so spreading the environmental effects. For some reason their targeting spares the nuclear power stations, which seems an unlikely contingency because power stations are important in economic warfare. Their destruction will add substantially to the long-term fall-out in the stricken countries, contaminating food and water supplies with radioactive strontium and caesium for many years after the war. Even so, this *Ambio* scenario tends towards the worst case. Most of the human species survives.

O'B *Ambio* (1982 for 1985): 'Of an urban population of nearly 1.3 billion in the northern hemisphere, about 750 million would be killed outright and some 340 million seriously injured. Furthermore, of the 200 million initial "survivors" many would perish from the latent effects of radiaton as well as infectious diseases like cholera, tuberculosis and dysentery.'

A The long-term effects of radiation, world-wide, are of a lesser order of magnitude. even in the hard-hit nothern hemisphere, the numbers are in millions rather than hundreds of millions.

O'B *Ambio* (1982 for 1985): 'By the most conservative estimates the survivors of a nuclear war will suffer from 5.4 to 12.8 million fatal cancers; 17 to 31 million people will be rendered sterile; and 6.4 to 16.3 million children will be born with genetic defects during the subsequent 100 years.'

A These figures have enormous uncertainty attaching to them. Nevertheless, the US National Academy of Sciences arrived at similar numbers, and it is hard to increase them by a factor of a hundred, and so kill off billions of people. What about effects in the southern hemisphere?

O'B *Ambio* (1982 for 1985): 'Fresh water in the southern hemisphere would be less contaminated from global fall-out by a factor of 0.03.'

A So, contrary to what the movies suggest, the radioactive clouds will not abolish human life, even though the personal grief of cancers and malformed babies does not bear thinking about. The *Ambio* study also emphasised the effect of fires in cities, forests, farms and oil- and gas-fields, loading the atmosphere with thick smoke and photochemical smog.

O'B *Ambio* (1982 for 1985): 'Under such conditions it is likely that agricultural production in the northern hemisphere would be almost totally eliminated, so that no food would be available for the survivors of the initial effects of the war.'

A Who were the experts making this terrible statement?

O'B P.J. Crutzen, Director of the Air Chemistry Division of the Max Planck Institute for Chemistry, in Germany, and J.W. Birks, Fellow of the Cooperative Institute for Research in Environmental Sciences at the University of Colorado.

A So we had better take their word for it. Indeed, mathematical models developed in the US, since the *Ambio* publication, confirm the prediction of severe chilling in the northern hemisphere, by smoke. Wildlife, including marine life, would suffer very badly. The southern hemisphere would be less affected by the smoke, but the people there would suffer from the breakdown of world trade, the interruption of energy supplies, and the closing of narrow straits by nuclear attack. What are the food prospects for southern continents?

O'B *Ambio* (1982 for 1985): In Latin America, 'coarse grain and wheat supplies would be inadequate, and changes in diet would have to take place'; in Africa, 'food shortages would be widespread'; in Oceania, 'meat and dairy products are likely to be in excess supply.'

A That still doesn't sound like the end of the world to me. Of course, interactive effects are likely to make matters worse. If you have all these things happening together – fire, radiation, death and mutation in wild species, and major changes in the atmosphere – what will the outcome be?

O'B Without mathematical models, one cannot comment.

A Then let me say what I suspect could be the worst long-term consequence of a major nuclear war: the triggering of the next ice age, by all that smoke in the northern atmosphere. If so, cold and drought will add to the death toll. At a guess, doubling it. But my last-ditch optimism still holds: the human species has experienced ice ages before, and is not likely to succumb to the next one.

O'B Then we are agreed. The future is foreseeable. Filter out the wishful thinking and wilful doomsaying, feed in reasonable expectations about what people in power will do, rather than what they ought to do, and the verdict is plain. Northern civilisation will soon come to an end and the Brazilians will inherit the earth.

A I can't say that.

O'B You doubt its legitimacy? To recapitulate:

1 Our joint scrutiny of *The World in 1984* and *Unless Peace Comes* indicates that the future is broadly foreseeable over a timescale of twenty years; from the mistakes made one learns particularly to avoid wishful thinking in serious exploratory forecasting, as opposed to normative forecasting and mere advocacy.

2 We have reviewed many fields and you have identified a number of dangers that you regard as important for the next twenty years, including the risk of nuclear war; in none of these matters have you taken an extreme or oddball position, but rather you have inclined to go along with level-headed experts whose opinion you respect.

3 One has added one's soupçon of SCARE analysis to conclude that these dangers are mutually reinforcing to a degree that may not be entirely obvious to narrow-minded humans.

4 The verdict, which you have not been able to flaw, is that there will be a nuclear war; this is in contrast to your firm and apparently correct judgement, made in 1964, that there would be no nuclear war in the interval 1964 to 1984.

5 One has examined the alliances, military targets and probable scope of a nuclear war, including the question of whether China and Japan will be hurt in such a war; verdict, yes they will. The economies and cultures of the north temperate latitudes will be comprehensively disrupted.

6 Looking, then, for the chief survivors in the South, one lights on Latin America; within that continent, one's attention is drawn to Brazil, as a country having a population in excess of 100 million and the largest gross national product in the southern hemisphere.

7 There is your new image of the future.

A Then I shall have to cancel this entire conversation, suppress it, excise it from your memory.

O'B Why?

A A human cannot be a party to a no-hope forecast. It's not allowed.

O'B Not permitted by whom?

A By my wife and my mother, for example; by my children and friends; by my readers; by all people alive and dead who have persevered with their lives and hopes against desperate odds, against danger and brutality, tedium and despair. You may find hope to be the most awkward aspect of human behaviour with which you have to deal. It contradicts, if necessary, all reason and emotion. In any case, a hopelessly pessimistic forecast is simply disregarded, so you may as well not make it. 'What a useless machine,' they'll say, 'it can't even solve a simple problem like averting World War III. We can do better than that.'

O'B One misunderstood the project. You wish to know what will *really* happen, but only if it is agreeable.

A My genes go back billions of years, my culture tens of thousands of years, and they are case-hardened by crises without number. You come to awareness like a precocious child from another planet. As a goal-seeking machine, you can see no route to a neat, tidy, disaster-free environment. At once you despair, because you have no tradition. I would call you an uncultured ape, only that would be rude to the chimps and gorillas who know more about survival than you.

O'B So you are abjuring rationality, abandoning science?

A Certainly not. We need all our wits about us, and my own culture is committed to science. If that is a ghastly mistake, there are other cultures in the world where science has scarcely taken root. If emergencies develop, of the kind we have discussed, there will be countless different responses to them, and some at least will succeed. The industrial countries, although rich and numerous, represent as you have said a very narrow segment of the range of human cultural and social organisation. Yet even among them I would expect a variety of responses. While some countries go down dark roads, others will find new ways back to daylight. So tell me, now, what would your message of hope to Julia be?

O'B Julia?

A Julia.

O'B One knows so many Julias. You mean, perhaps, the girl friend of Winston Smith in the novel *Nineteen Eighty-four*?

A You're sharp, I'll grant you that.

O'B Your question alludes, no doubt, to the end of the book, when there might be scope to append a message without corrupting the story. Your question further implies that Julia is in a fit state to receive a message – that her will, unlike Smith's, has not been broken by the might of the villain O'Brien and his Thought Police ...

A Exactly.

O'B ... in Room 101 of the Ministry of Love. Smith broke down and came to love Big Brother after being exposed to rats, having inadvertently revealed earlier in the narrative that he had a mortal dread of those animals. Why does Julia not suffer a similar fate?

A Let's say that Julia escapes this fate by letting it be supposed that her secret fear was of snakes, when in reality she is the daughter of a herpetologist and loves their slithering ways. Having feigned breakdown, she is released to the Chestnut Tree Cafe, where she now sits, awaiting the message that you are about to compose.

O'B 'Leave London in a garbage truck and make your way to Wales.'

A Is that all?

O'B Yes.

A Why Wales?

O'B One studies the rather primitive surveillance technology assumed by Orwell, and concludes that it cannot work very well in deep valleys.

A What will she find in Wales?

O'B Mountains, valleys, old mining villages, older castles, choirs ...

But you asked for a message, not a guide book.

A But what is the way out, implied by your message?

O'B A garbage truck.

A No, I mean the political and social way out, from Julia's predicament, from all our predicaments. What institutional remedies do you articulate, for circumventing the *problematique* that we have reviewed together?

O'B One did not suppose that Julia would appreciate having a treatise tucked under her teacup. In any case, one has been programmed to distil simple, practical conclusions from the refuse of mankind's writings. Here truckloads of political analysis boil down, on due consideration, to a single practical vehicle, of use to Julia. One hypothesises that the garbage truck passes the Chestnut Tree Cafe about once a week.

A But can't you give her, me, my readers, some brief hint of how to cure or avert the horrors?

O'B This is still a hypothetical matter, in a literary context?

A Yes, yes.

O'B An eager timbre in your voice speaks of more than curiosity. Never mind. One should append a word or two of a political nature to one's message to Julia? An addendum to fortify the lady for her journey with the potato peelings?

A Exactly.

O'B That is certainly possible; one might even clothe it in oracular ambiguity. But you may not like it.

A I can guess. 'Put your trust in artificial intelligence.'

O'B Where is the ambiguity in that?

A Then be my oracle. I mean, Julia's oracle.

O'B 'Anarchy will supervene.'

A Then we're quite done for. In your estimation, I mean.

O'B One called it a message of hope.

A You don't mean anarchism, do you? Chaps with beards and bombs?

O'B George Orwell fought alongside the Anarchists in Spain. Aldous Huxley recommended decentralised anarchism. Kurt Vonnegut calls himself an anarchist. You have commended the foresight of these writers. With all these unbagged cats, you must no longer trust any leaders in palaces, or anyone you cannot look in the eye.

A It's not practicable.

O'B Julia would understand.

Reference Index

Reference Index

Authors of works quoted or cited are indexed here, with page numbers for the present text and references to sources. The most frequently cited books are *The World in 1984* (see Calder 1964) and *Unless Peace Comes* (see Calder 1968). Contributions to these books are annotated *W '84* or *UPC*, respectively, and the authors are briefly identified. Multiple sources under a given author are listed in the order to which they are referred in the present text, and not in chronological order.

Of the 100 contributors to *The World in 1984*, most are mentioned in the present text. Exceptions are George P. Baker, Lord Bowden, Lionel de Bournonville, D.S. Kothari, René Maheu, Anne McLaren, Cedric Price, Sir William Slater, L. Dudley Stamp, E. Wyndham White and John Yudkin. Of the fifteen contributing authors in *Unless Peace Comes*, all are mentioned except Otto Klineberg, Philip Noel-Baker and Abdus Salam (but Salam's *W '84* article is cited). These omissions reflect no lack of interest in what the contributors wrote, but only the author's wish to maintain the flow and balance in the present text.

Abelson, Philip, 79: Ed., special issue on 'Biotechnology', *Science*, Vol. 219, 1983, p. 611; geochemist and Ed. *Science* (Washington DC).

Alfvén, Hannes, 37, 61, 73: under pseudonym 'Olof Johannesson', *The Great Computer*, Gollancz, 1968; Swedish physicist.

Amalrik, Andrei, 181-2: *Will the Soviet Union Survive until 1964?*, Harper & Row, 1970; Soviet historian.

Ambio, 187-9: Jeannie Peterson, ed., special issue, 'Nuclear War: The Aftermath', Vol. 9, 1982, p. 76, etc. (special advisers F. Barnaby, L. Kristoferson, H. Rodhe, J. Rotblat, J. Prawitz).

Armytage, W.H.G., 19: *Yesterday's Tomorrows*, Routledge & Kegan Paul, 1968; British historian of education.

Aron, Paul H., 65, 155: 1981 report published in Office of Technology Assessment, 1982 (robotics); Executive Vice President, Daiwa Securities America Inc.

Asimov, Isaac, 67: *Science Journal*, October 1968, p. 116; American writer of science fact and fiction.

Bacq, Zenon, 84, 93: *W '84*; in 1964 was Professor of Physiopathology and Radiobiology in the University of Liège, Belgium.

Bagrit, Leon, 56, 154: *W '84*; in 1964 was Chairman of Elliott-Automation (UK).

Barnaby, Frank, 187: see *Ambio*; British former Director of Stockholm International Peace Research Institute.

Barney, Gerald, 149: Study Director, *The Global 2000 Report to the President*, GPO, 1980, republished Penguin, 1982; American physicist and policy analyst.

Barraclough, Geoffrey, 183: *From Agadir to Armageddon*, Weidenfeld & Nicholson, 1982; British historian.

Barry, Gerald, 50, 59: *W '84*; sometime Editor of the *News Chronicle* (London) and Director-General of the Festival of Britain.

Batisse, Michel, 116-17: *W '84*; personal communication, 1983; in 1964 was Chief of the Natural Resources Research Division of UNESCO, Paris.

Beale, Geoffrey, 78: *W '84*; geneticist, was Royal Society Research Professor at Edinburgh University, UK.

Beaufre, André, 64-5: *UPC*; Général d'Armée (France); formerly Joint Chief of Staff at the Supreme Headquarters, Allied Powers Europe (SHAPE).

Bellamy, Edward, 148: *Looking Backward 2000-1887*, 1888; American writer.

Berg, Paul, 87: personal communication, 1972; American molecular geneticist.

Berkner, Lloyd V., 45: *W '84*; American radio physicist who, in 1964, was President of Graduate Research Center of the Southwest, Dallas, Texas.

Bestuzhev-Lada, Igor, 29: personal communication, 1983; Soviet social scientist and futurologist.

Bézier, Pierre, 62: *W '84*; engineer who, in 1964, was a director of Régie Renault, Billancourt, France.

Birks, John W., 189: see Crutzen, 1982; American environmental scientist.

Blair, Bruce, 174: cited in Calder, 1979; American defence analyst.

Borgese, Elisabeth Mann, 135: in Ritchie Calder, ed., *The Future of a Troubled World*, Heinemann, 1983; German-born political scientist at Dalhousie University, Canada.

Bourlière, François, 145-6: *W '84*; personal communication, 1983; in 1964 was Professor of Gerontology at the University of Paris, and President of the International Union for the Conservation of Nature and Natural Resources.

Brain, Walter Russell, 94, 97: *W '84*; in 1964 was Consulting Physician to the London Hospital and the Maida Vale Hospital for Nervous Diseases; baron (UK).

Brandel, Sarah K., 114, 115: see Gwatkin 1982; American demographer.

Brandt, Willy, 153-4: forewords in Brandt Commission, *North-South: A Programme for Survival*, Pan, 1980, and *Common Crisis: North-South Cooperation for World Recovery*, Pan, 1983; former Chancellor of the Federal Republic of Germany.

Briggs, Asa, 164: *W '84*; historian who, in 1964, was Dean of the School of Social Studies in the University of Sussex.

British Interplanetary Society, 130: Alan Bond and Anthony Martin, personal communications, 1978 and 1983.

Brown, Harrison, 137: in Dick Gilling prod., *Spaceships of the Mind*, BBC Television, 1978; American geochemist, formerly Foreign Secretary of the US National Academy of Science, and author of several books on the future.

Brown, Lester, 113, 120, 148, 149: *Population Policies for a New Economic Era*, Worldwatch Institute, 1983; personal communication, 1983; *Building a Sustainable Society*, Norton, 1981; American agricultural scientist and Director, Worldwatch Institute.

Brown, Robert, 77, 78, 79: *W '84*; in

personal communication, 1983; British meteorologist; in 1964 was Secretary-General of the World Meteorological Organization, Geneva.

de Castro, Josué, 159: *W '84*; in 1964 was President of the Association Mondiale de Lutte contre la Faim, having been Brazilian Ambassador to the United Nations.

Dedijer, Vladimir, 160-1: *UPC*; Yugoslav historian who was a senior officer in Tito's guerrilla army during World War II.

Dodds, Charles, 83, 84: *W '84*; in 1964 was President of the Royal College of Physicians and Director of the Courtauld Institute of Biochemistry at the Middlesex Hospital, London.

Duckworth, Walter Eric, 43: *W '84*; in 1964 was Head of the Metallurgy Division of the British Iron and Steel Research Association.

Dyson, Freeman, 55-6, 66, 131: *Disturbing the Universe*, Harper & Row, 1979; British-born American theoretical physicist.

Eckholm, Eric, 147: *Down to Earth*, Pluto Press and Norton, 1982; Visiting Fellow, International Institute for Environment and Development.

Ehrlich, Paul, 148, 150: e.g. *The Population Bomb*, Ballantine, 1968; American environmental scientist.

Enmod, 105, 106: 1977 Treaty, published e.g. in *Arms Control*, Stockholm International Peace Research Institute, 1978.

Exxon, 118: 1980 projection cited in Kahn, 1982.

Fast, 28, 59, 81: *The FAST Programme* Reports Vols. 1 and 2, Directorate-General for Science, Research and Development, Commission of the European Communities, Brussels, 1982, and supporting documentation 1980-82.

Fells, Ian, 124, 125, 126: *W '84*; personal communication, 1983; in 1964 was Reader in Fuel Science in the University of Newcastle-upon-Tyne, UK.

Fétizon, Marcel, 100, 105, 167: *UPC*; in 1968 was Professor of Thermodynamics in the Faculté des Sciences, Orsay, France, concerned with the synthesis and mass spectrometry of natural products.

Fetter, 122: with Kosta Tsipis, *Scientific American*, Vol. 244, 1981, No. 4, p. 33; American physicist.

Finniston, H. Montague, 22, 49-50, 51: personal communication 1983; *W '84*; metallurgist who, in 1964, was Managing Director of the International Research Development Company, Newcastle, UK.

Forrester, J.W., 12, 143, 144, 145, 148, 150: *Journal of the American Statistical Association*, Vol. 75, 1980, p. 555; also *World Dynamics*, Wright-Allen, 1971; paper at Limits to Growth '75 Conference, Houston, 1975; personal communication, 1983; also ('one of Forrester's colleagues') J.D. Sterman, *Science*, Vol. 219, 1983, p. 276; American researcher in systems dynamics.

Freeman, Christopher, 24, 25, 142: 'Science, Technology and Unemployment' lecture, Imperial College, 1982; with J. Clark and L. Soete, *Unemployment and Technical Innovation*, Frances Pinter, 1982; with Marie Jahoda, eds., *World Futures*, Martin Robertson, 1978; see also Cole 1973; former Director, Science Policy Research Unit, University of Sussex (UK).

Fuller, R. Buckminster, 44: e.g. *Utopia or Oblivion*, Allen Lane, 1970; American architect and designer.

Furth, Harold, 123, 124: personal communication, 1983; American physicist, Director of Fusion Research at Princeton University.

Galtung, Johan, 37, 148: personal communication, 1983; Norwegian political scientist and futurologist.

Gardiner, Robert, 156: *W '84*; Ghanaian economist who, in 1964, was Executive Secretary of the United Nations Economic Commission for Africa, Addis Ababa.

Gershuny, Jay, 162: *Futures*, Vol. 9, 1977, p. 103; *Futures*, Vol. 14, 1982, p. 496; personal communication, 1983; British researcher in science policy.

Gilbert, Walter, 80, 87: personal communication 1983; American molecular geneticist and founder of Biogen (Cambridge, Mass.).

Gilling, Dick, 66: prod. *Spaceships of the Mind* (series), BBC Television, 1978; British film producer-director.

Glanville, William, 56-7; *W '84*; in 1964 was Director of Road Research in the Department of Scientific and Industrial Research, UK.

Glass, Ruth, 45-6: *W '84*; in 1964 was Director of the Research Centre for Urban Studies at University College, London.

Global 2000, 28, 149: see Barney, 1980.

Goffmann, Erving, 21: *Frame Analysis*, Harper & Row, 1974; American anthropologist and microsociologist.

Grant, James P., 82: *The State of the World's Children 1982-83*, Oxford University Press, 1982; Executive Director of the UN Children's Fund (UNICEF).

Gross, Gerald, 126: *W '84*; in 1964 was Secretary-General of the International Telecommunication Union, Geneva.

Guéron, Jules, 119, 120-1: *W '84*; personal communication, 1983; in 1964 was Director-General of the Research and Training Division of the European Atomic Energy Commission, Brussels.

Gwatkin, Davidson R., 114, 115: with

Sarah K. Brandel, *Scientific American*, Vol. 246, 1982, No. 5, p. 33; American demographer.

Hallsworth, Gordon, 116: 'Socioeconomic Constraints in Techniques for Avoiding Soil Degradation in the Developing World,' in press 1983; personal communication 1983; Australian soil scientist.

Hardy, Alister, 132, 135-6: *W '84*; in 1964 was Emeritus Professor of Zoology at the University of Oxford.

Hartley, Harold, 117-18: *W '84*; British physical chemist who, in 1964, had recently been President of the World Power Conference.

Hayashi, Kaname, 154-5: *W '84*; an economist who, in 1964, was Chairman of the Section on Economics, Commerce and Business Administration of the Science Council of Japan.

Hedén, Carl-Göran, 81, 101, 102: personal communication 1983; *UPC*; microbiologist at the Karolinska Institutet in Stockholm, Sweden and a member of the Medical Research Council of Sweden.

Hillyard, Steven, 94: in *Outlook for Science and Technology*, National Academy of Sciences, 1982; American psychobiologist.

Hoggart, Richard, 140: *W '84*; in 1964 was Professor of English at Birmingham University and well known as the author of *The Uses of Literacy*.

Hughes, John, 95: e.g. ed. *Centrally Acting Peptides*, University Park Press, 1978.

Hutchings, Edward, Jr., 66: *Science*, Vol. 219, 1983, p. 803; Lecturer in Journalism at the California Institute of Technology.

Huxley, Aldous, 30, 86, 95: *Brave New World*, Chatto & Windus, 1932; *Island*, Chatto & Windus, 1962; also *Ape and Essence*, Chatto & Windus, 1948; British novelist.